great demonstration convinced Snow that man-kind could fight injustice by peaceful means. While America mobilized for World War II, he de-nounced all war as evil, remaining a committed pacifist until his death in 1955. Shortly before he died, Snow published an autobiographical mem-oir, *From Missouri*, in which he affirmed his opti-mistic belief that people could peacefully change the world.

This biography places Snow in the context of his place and time, revealing a unique individual who agonized over racial and economic oppression and environmental degradation. Snow lived, worked, and pondered the connections among these issues in a small rural corner of Missouri, but he thought in global terms. In a new millennium, with the civil rights movement and a series of wars to inform us, these issues still demand our attention today. Well-crafted and highly readable, *Thad Snow* provides an astounding assessment of an agricultural entrepre-neur transformed into a social critic and an activist.

Photo by Michael Grace

About the Author

Bonnie Stepenoff is Professor of History at Southeast Missouri State University in Cape Girardeau. She is the author of *Their Fathers' Daughters: Silk Mill Workers in Northeastern Pennsylvania, 1880–1960*.

Thad Snow

Missouri Biography Series

William E. Foley, Editor

Thad Snow

A Life of Social Reform

in the Missouri Bootheel

Bonnie Stepenoff

University of Missouri Press
Columbia and London

Missouri Center
for the Book

🙚🙚🙚

Missouri Authors
Collection

Cataloging-in-Publication data available from the
Library of Congress
ISBN 0-8262-1496-7

∞™ This paper meets the requirements of the
American National Standard for Permanence of Paper
for Printed Library Materials, Z39.48, 1984.

Designer: Jennifer Cropp
Typesetter: Crane Composition, Inc.
Printer and binder: Thomson-Shore, Inc.
Typefaces: Palatino and Wade Sans Light

Frontispiece: Farm building at Snow's Corner,
Mississippi County, Missouri. *Photograph by Bonnie
Stepenoff.*

To Jerry, Samantha, and Hannah,
and to Thad Snow,

"... who says he found central Indiana, where he was born and reared, too tame and therefore settled in southeast Missouri, cleared a thousand acres, farmed it and made it pay. Along the way he fought the Mississippi in flood and Army engineers on flood control. His book *From Missouri* is a personal story and a sociological study of the land and the people."

 —*Indianapolis News*, November 6, 1954

Contents

Preface   ix

Acknowledgments   xv

Introduction   1

1. Snow's Corner   9

2. The Big-Eye   23

3. Flood Culture   39

4. King Cotton   53

5. Out on Mr. Snow's Farm   71

6. The Great Roadside Demonstration   89

7. Bootheel Planter   115

8. Missouri Pacifist   126

9. Ozarks Retreat   138

10. From Missouri   155

Epilogue   163

Selected Bibliography   165

Index   175

Preface

Well, he was kind of an odd kind of a guy.

> —Nellie Feezor Stallings, Charleston, Missouri

He was thinking faster than most people. That was what his problem was. He was thinking ahead of his times.

> —Hunter Rafferty, Wyatt, Missouri

I got acquainted with Thad through my father. They got along pretty well. But I wasn't sure my father really knew how to take Thad. He was quite a bit different from most hillbillies.

> —George Burrows

On July 10, 1998, I went to Van Buren, a small town on the Current River in the Missouri Ozarks, searching for information about Thad Snow. Alan Turley, publisher of the local newspaper, the *Current Local,* said he had no files on Snow, but suggested I go down the street and talk to George Burrows.[1] Burrows's father, Emmett Russell "Rip" Burrows, the postmaster at Van Buren, was a close friend of Snow's in the 1950s, when he lived at the Rose Cliff Hotel.

I found George Burrows working with a table saw in the driveway behind his house. He turned off the machinery and greeted me

---

1. George Burrows, a retired schoolteacher, died on July 4, 2002, at the age of eighty-two. According to his obituary in the *Van Buren Current Local,* July 11, 2002, "He was a naturalist who enjoyed farming, hunting and fishing, and was an accomplished guitar player."

with a smile. When I told him what I wanted, he said he was happy to talk about Snow. In fact, he said there was nothing he liked to do better than to talk about him.

When I visited Burrows again on a winter day in 1998, he invited me inside, and we sat and talked in his living room. He told me he had trouble sleeping. When he woke up in the middle of the night, he would read or play his electric guitar. He had lived in the Ozarks most of his life. His father's mother was born in 1857 in a two-room log house on Pike Creek. She was named after her father, Samuel Cusick. Everyone called her "Aunt Cuse." Her grandson George was born in the same log house, about two miles from Van Buren.

Burrows went to Arkansas State College at Jonesboro in 1939, when the entire student body was housed in one building. He also studied for two years at the University of Missouri in Columbia before serving in the medical corps during World War II. He finally earned his degree at Jonesboro and taught school for several years in St. Louis County. He returned to the Ozarks and worked as a naturalist at Big Spring State Park.

"There was a little stone building," he said. "That was the museum. I was supposed to catch snakes, which I did—put snakes on display, rattlesnakes. Once I had a little deer. People brought me little baby coons. I took care of them and kept them there. Once I had a coyote. There was a little fish tank out front. I kept water snakes and turtles in that. Part of the job was to take people on walks. They called it nature tours. You'd try to answer as many questions as you could."

He didn't remember exactly when Snow appeared in Van Buren. "Along in the early fifties," he thought. Leafing through a scrapbook, he found a photograph of himself and Thad Snow at Big Spring.

Burrows was a young man when Snow lived in retirement at the Rose Cliff, where he kept a well-stocked bar in his rooms and "liked a toddy in the afternoon." Snow's rooms were in the basement, but they gave him a view of the Current River. "He looked out," said Burrows, "the way it was built—he had windows—he could see. He wasn't buried."[2] Snow spent a great deal of his time there, writing his memoirs.

According to Burrows, Snow was a liberal, a socialist, but not a

2. George Burrows, interview with the author, Van Buren, Mo., December 15, 1998.

communist, as some people believed. During his years in the Ozarks, he lived quietly and was not involved in politics. He kept a room in the house at Snow's Corner, near Charleston, Missouri, where he had farmed for forty years, but he spent most of his time in the Ozarks.

Snow came to Van Buren to escape from memories of a terrible tragedy. In August 1948, his son-in-law John Hartwell Thompson shot and killed his wife, Priscilla, their daughter, Ann, and his sister-in-law Emily, before committing suicide.[3] On one summer night, Snow lost his eldest and youngest daughters and his granddaughter. Snow's only surviving daughter, Fannie, and her husband, Bob Delaney, moved to Snow's Corner and took over the farm.

As Burrows recalled, Fannie and her husband used to come to Van Buren and visit, along with a black housekeeper named Daisy. Burrows said Fannie Delaney had a paralyzed arm, which had been injured by an airplane propeller. She had refused to allow an amputation. She was stubborn, said Burrows, "a lot like her father."[4]

Fannie Delaney remembered that her father "held court in that little room at the Rose Cliff Hotel,"[5] welcoming out-of-town visitors who wanted to talk about the great issues of the time. The writer and outdoorsman Leonard Hall liked to sit with him by a campfire and philosophize. Reporters, editors, and cartoonists came from St. Louis to chat and watch the river go by.

Burrows remembered one day when he was at the post office; "Thad came in and said to my father, 'Dan Fitzpatrick [the cartoonist] is down at T. O. Wright's place down on the river, down at Gooseneck. Let's go down and see him.' And it was just relaxing and conversation. Dan was sitting there, nobody paying any particular attention to him. After a bit he walked over to my father and handed him a caricature. Amazing, amazing."

Snow stood out as an eccentric in Van Buren. "He was here during the Korean conflict," said Burrows, "and you know it wasn't popular back in those days to be opposed to any nationally sponsored event. He was very strongly opposed to the Korean War. People

3. *St. Louis Post-Dispatch*, January 16, 1955.
4. Burrows interview, December 15, 1998.
5. Fannie Snow Delaney, interview with the author, Snow's Corner, Mississippi County, Mo., August 16, 1999.

around here didn't know how to take it. He just wasn't a popular person. He just didn't make a lot of friends. Didn't want them, I guess. He was just different from us, vastly different."

He and Burrows shared a passion for hunting. Although Snow was in failing health—"couldn't follow a dog anymore"—he "still liked to hunt deer in the Ozarks and duck on the Illinois side of the river." A few times, Burrows took him quail hunting in the Ozarks: "He couldn't walk too well. I was the only person in Van Buren who ever got an invite to go over to Thad Snow's place and duck hunt. . . . They owned a portion of an oxbow lake—Thad and about a dozen of his friends. They had permanent blinds. And the next nearest blind was so far that you could barely hear it when they fired. Now that's just how big it was. It was probably the best place in the United States to go duck hunting. That was when there were lots and lots of ducks."

Burrows was saddened because there were not so many ducks anymore and very few quail in the Ozarks: "I think it's the change in farming practices. It's fescue and cattle. When I was hunting quail, it was mixed farming, corn and cattle."

He believed that Snow shared his concern for the natural world. "Oh, yes, he was a conservationist. No question about it. An environmentalist, though that word was barely known. If he were alive now, he would be, as I am, a member of the Sierra Club."[6]

Snow's concern for the environment came late in his life. When he first came to Missouri in 1910, he cleared and drained about a thousand acres of swampland in Mississippi County, one of six counties in the state's southeastern Bootheel. By the early 1920s he had turned those acres into a profitable cotton plantation, worked primarily by African American sharecroppers.

He had mixed feelings about the sharecropping system. In the 1930s, when the Great Depression brought misery to farm laborers, he became an outspoken critic of the economic and social systems that denied justice to people who were poor and black. He encouraged the workers on his farm to join a union, and he supported the sharecroppers' roadside demonstrations in 1939. When he came to Van Buren, there were still sharecroppers living on his farm. Unlike

6. Burrows interview, December 15, 1998.

most Bootheel farmers, he resisted the trend toward dismissing share-croppers and employing transient day laborers. In the 1950s, his daughter and son-in-law mechanized the farming operation.

More than fifty years later, his daughter remembered, with some exaggeration and some truth, that "Everybody in Mississippi County hated him, because he held the view that blacks were peo-ple, too."[7] Hunter Rafferty of Wyatt, Missouri, remembered that Snow was "very active on the part of the sharecroppers." Most of his white, middle-class neighbors, however, bore him no ill will. According to Rafferty, "We didn't fall out with him."[8] In his mem-oir, *From Missouri*, published in 1954, shortly before his death, Snow returned again and again to the subject of the sharecroppers, their suffering, and their heroic Depression-era protest.

At the end of his life, he questioned his own motives in turning a tract of wild swampland into a cotton plantation. He regretted the loss of trees and wildlife and perceived a connection between the exploitation of the land and the oppression of the people.

Snow impressed Burrows so much that he named his son after him. "He was a very strong environmental person," said Burrows, "strongly opposed to many of the things big businessmen do. Oh, he admired Gandhi. At his instigation I read a couple of books—one that Gandhi had written on peaceful resistance. Another one, if I could find it, I'd read it again, [Thorstein Veblen's] *The Theory of the Leisure Class.*"

When Snow died in 1955, the *Van Buren Current Local* carried a brief obituary, which stated that "he lived a colorful life, and his views on religious, social, and political questions, while not always orthodox, were strong and made him both friends and enemies. He will be missed but long remembered by the friends he made here."[9]

At the end of the twentieth century, Burrows remembered him vividly, but Burrows was a generation younger than Snow. Snow's contemporaries were gone. The sharecroppers who worked on his farm had dispersed or passed away. In 2002, Jim Robinson, a black deputy sheriff in Mississippi County, said that he was "still a little

7. Fannie Snow Delaney, interview with the author, Snow's Corner, Missis-sippi County, Mo., March 19, 2001.
8. Hunter Rafferty, telephone interview with the author, May 28, 2002.
9. *Van Buren Current Local,* January 20, 1955.

boy" in 1939, when the Missouri sharecroppers went out on strike. He said, "All those folks are gone."[10]

This book is an attempt to reconstruct, from books, newspaper articles, manuscripts, government records, and oral sources, the world in which Snow lived his colorful life. Some of the anticipated sources proved difficult or impossible to find. For instance, an FBI case file on Snow was destroyed in 1991. From available evidence, however, it was possible to piece together the story of a man trying to come to grips with economic, social, and environmental forces in which he played a small and not always comfortable role. As his friend Hunter Rafferty said, "He was thinking faster than most people."[11] But his thinking did not always keep up with the complexities of the world in which he lived.

10. James Robinson, telephone interview with the author, September 10, 2002.
11. Rafferty interview, May 28, 2002.

Acknowledgments

I would like to thank Fannie Snow Delaney, George Burrows, Nellie Feezor Stallings, Shirley Whitfield Farmer, and Julia Cooper Warren for sharing their personal memories of Thad Snow, Owen Whitfield, Emily Snow, the Bootheel, and the Ozarks in the momentous years before 1955. Mr. Burrows gave me a glimpse of how it might have been to sit on a riverbank discussing the fate of the world with his old friend Thad. Mrs. Stallings remembered growing up on Snow's farm. Mrs. Farmer recalled how her father, Owen Whitfield, stood up to demand justice for the sharecroppers. Mrs. Warren remembered laughing and having a good time with her best friend, Emily Snow. Mrs. Delaney, Thad Snow's only surviving daughter, reaffirmed her father's belief that justice was not only desirable, but necessary; not only imaginable, but possible.

Renae Farris diligently searched the papers of Governor Lloyd Crow Stark and uncovered important information. Will Sarvis guided me through collections of oral interviews and manuscripts at the State Historical Society of Missouri. Brian Driscoll did thorough and helpful research in deeds and land records. Thanks are due to reference librarians and archivists at Clara Drinkwater Newnam Library in Charleston, Missouri; Kent Library at Southeast Missouri State University; Western Historical Manuscripts Collection at the University of Missouri–Columbia; and the Library of Congress. I owe special thanks to Ms. Hazel Williams of Charleston for finding Thad Snow's probate records.

Several scholars and editors offered advice and commentary on various drafts of the book. Dr. Gary Kremer read and reread the manuscript and offered invaluable guidance. Professor William Foley made incisive comments on the text. Dr. Susan Flader encouraged me to pursue the research on Thad Snow and the Bootheel. Earlier versions of portions of this manuscript have been published in *The Missouri Historical Review* and *The Red River Valley Historical Journal.* Many thanks to Julianna Schroeder for her inspired editing of the book.

The Richard S. Brownlee Fund of the State Historical Society of Missouri and the Grants and Research Funding Committee of Southeast Missouri State University generously provided financial assistance for my research.

My husband, Jerry, and my daughters, Samantha and Hannah, have made it all worthwhile.

Thad Snow

Introduction

The understanding of the present is ever more difficult and obscure than the study of the past; and as for looking into the future, the human mind is notably lacking in the faculty for doing that.

—Thad Snow

As a young man in the 1890s, Thad Snow feared that he did not live in interesting times. He was wrong. In the early 1930s, he reflected that most of his "youthful hopes and ambitions have faded away in the struggle with the insistent realities of life." But one of his early wishes, the one he "least expected to see fulfilled," did come true. In middle and old age, with some excitement and much regret, he witnessed "real history in the making."[1]

In his prime, he stood six feet tall and weighed two hundred pounds, a man of great strength and stamina, capable of putting in long hours of farm labor.[2] Although he chose farming as his occupation, his talents included playing the piano, writing humorous essays and polemics, and debating the issues of the day. He smoked Kool cigarettes and drank Manhattans.[3] His children's friends marveled at the library he maintained in his house at a rural crossroads

1. Thad Snow, "When Traders Rule, Ruin Follows," 13.
2. *St. Louis Post-Dispatch,* January 16, 1955.
3. Julia Cooper Warren, interview with the author, Charleston, Mo., November 7, 2002.

that had become known as "Snow's Corner."[4] Perhaps his greatest achievement was observing the life around him and recognizing history as it unfolded in his own backyard.

Acquaintances viewed him as a paradoxical character; he often expressed unorthodox opinions but chatted amiably with all kinds of people on all kinds of subjects, from mules and hunting dogs to philosophy and politics.[5] Hunter Rafferty noted that Snow liked to have a drink, even early in the morning.[6] Jennie Cooper, the daughter of a Baptist minister, remembered seeing Snow "in his seersucker suit and straw hat, sitting with my father and philosophizing." Amusingly, she said, "We considered Thad Snow to be the local atheist-communist-socialist, but he and my dad were great friends."[7]

Born in 1881, Snow spent his childhood in Greenfield, Indiana. Early in the twentieth century, he began farming on land owned by his father. On March 19, 1903, he married Bess Jackson, the twenty-two-year-old daughter of Solomon and Adah Jackson of Greenfield. Bessie became the mother of two children, Hal (born in 1905) and Priscilla (born in 1907). As a young and ambitious farmer, Thad Snow could not foresee that Bessie would die violently in a fall from a horse in 1915 and that his own life would follow a strange and twisting road.[8]

Convinced that farming was his life's work, he went searching for undeveloped land along the Mississippi River during the lumber boom in 1910. The following January, he purchased a tract of nearly eleven hundred acres (1081.81 acres) near Charleston in Mississippi County, Missouri.[9] Less than a third of the land had been cleared, and the rest was wooded wetland. According to Snow, there were six houses on the property, including one old log

4. Jennie Cooper, telephone interview with the author, September 23, 2002.

5. *Charleston (Mo.) Enterprise-Courier,* January 20, 1955.

6. Rafferty interview, May 28, 2002.

7. Jennie Cooper interview, September 23, 2002.

8. *St. Louis Post-Dispatch,* January 16, 1955; Nunnelee Funeral Home, Death Records, Mississippi County, Missouri, 1910-1930, compiled by the Mississippi County Genealogical Society, Charleston, Mo., 30.

9. Warranty deed, January 23, 1911, Mississippi County, Missouri, deed book 63, p. 581. Snow's property was located in township 26N, range 16E and range 17E.

house, one four-room house, and four small tenant houses.[10] He paid fifty-one thousand dollars for this property, a real bargain, since the average value per acre in the county was sixty-four dollars in 1910. Snow's purchase seemed a shrewd one, because real estate values doubled in Mississippi County between 1910 and 1920.[11]

In the 1920s, he began a new family with his second wife, the former Lila Simpson, daughter of John L. and Fannie Simpson of Charleston. The forty-year-old Snow married thirty-four-year-old Lila on March 11, 1921.[12] Their daughter Lena Frances (Fannie) was born in 1922, and her sister Emily came along about four years later. Fannie remembered that her father liked to read to his daughters and that he taught them to ride and shoot. Her favorite pastime was riding, but she did not become an expert hunter. As she recalled, her father did not believe she was a very good shot.[13]

For nearly forty years, Thad Snow lived and worked at Snow's Corner while the landscape around him changed from thickly wooded wetland to flat treeless fields crisscrossed by drainage ditches and highways. He drained his land, hired woodsmen and their families to cut down trees, lobbied for better roads, and helped "reclaim" Missouri's wild Bootheel for agricultural production. But the process and his part in it troubled his conscience. He was never sure he had done the right thing.

At the end of his life, he recalled his move to Missouri as the act of a modern-day pioneer. Like many pioneers before him, he went looking for a challenge, and he was admittedly ambitious. In part, he saw the chance to make a profit, but in a less materialistic sense, he also wanted to leave his imprint on the land. As he aged, he became

10. Thad Snow, *From Missouri*, 96.

11. He had a financial partner in this venture. Anna D. Cooper, the widow of Frank Cooper, of Henry County, Indiana, owned a half interest in the property until December 1926, when Thad Snow bought her share for twenty thousand dollars. This transaction was recorded in Mississippi County, Missouri, deed book 98, p. 505. At that time the property was subject to a mortgage of thirty-eight thousand dollars from Phoenix Mutual Life Insurance Company. See deed book 59, p. 476; and Thomas J. Pressly and William Scofield, eds., *Farm Real Estate Values in the United States by Counties, 1850–1959*.

12. Mississippi County Marriage Licenses, book 13, p. 415. Genealogical information on Lila Simpson is available on Ancestry.com.

13. Delaney interview, March 19, 2001.

increasingly aware of the irony of the frontier. Initially, he migrated to the Bootheel because it was a wild place, but little by little he tried to recreate it in the image of the more settled place he left behind. History intervened, however, and southeastern Missouri developed in ways he never could have imagined.

A land developer in Charleston assured him that, because of a new levee system, the county would never flood again. Surely, Snow did not believe the promise. He knew there would be overflows, but the power of the Mississippi River surprised and chastened him. Floods in 1912 and 1913 served notice that the river had little respect for human efforts to keep it in check. The disastrous flood of 1927 inspired massive public works by the U.S. Army Corps of Engineers. Despite these efforts, and in part because of them, the flood of 1937 brought terrible suffering to people, especially farm laborers, in the Bootheel. In the spring, after the flood, Snow lost his second wife, Lila, to a long illness. By this time, Snow had begun writing letters and essays about the suffering of farmworkers. His wife's death shook him mentally and physically. Although he left the Bootheel for extended periods, traveling to regain his health, in the late 1930s he spent increasing amounts of time thinking and writing about the region's troubles.[14]

Of all the disasters that befell the Bootheel, cotton was the most fateful. As soon as developers cleared and drained the lowlands in the early twentieth century, cotton planters, ginners, tenant farmers, and sharecroppers migrated into the region. Beginning in 1923, Snow planted some of his acreage in cotton. By the early 1930s, he had twenty sharecroppers and their families living on his land. Although he profited from cotton production, he realized that doing so required working sharecroppers and day laborers as hard as he could and paying them as little as possible.[15] When cotton prices fell after 1929, the situation became appalling, and Snow became troubled.

The Great Depression convinced Snow that, in spite of his youthful prediction, he and his life had become part of history. By the late 1920s the cotton economy crashed and land values sagged. Snow

---

14. Snow, *From Missouri*, 93, 220; *Poplar Bluff (Mo.) Daily American Republic,* January 17, 1955.
15. *St. Louis Post-Dispatch,* May 2, 1937.

found himself in financial difficulties.[16] Like most cotton growers, he welcomed New Deal programs that compensated farmers for plowing up cotton and taking fields out of production. But these well-intentioned subsidies increased the misery of tenant farmers and sharecroppers. With government incentives, many landowners reduced production drastically and evicted tenants from the cotton fields.

Snow wrote impassioned letters to the U.S. Department of Agriculture and to the *St. Louis Post-Dispatch*, calling attention to the plight of dispossessed farm laborers in the Bootheel. He believed New Deal administrators meant well but that they failed to truly understand the situation in southeastern Missouri. He also believed that some planters deliberately cheated their croppers out of benefits due them from the government. Misguided policies, bungling, and cheating caused more than just hardship. In a 1934 letter to a government official, Snow described the treatment of sharecroppers as "an injustice and betrayal that burns deep into their souls."[17]

In 1936, John L. Handcox, an organizer for the Southern Tenant Farmers' Union (STFU), came to Mississippi County and received a warm welcome from Snow. The STFU, which originated in Arkansas in 1934, was a biracial labor union with socialist connections, a Christian base, a crusading spirit, and a straightforward mission to win fair treatment and dignity for farm laborers. Handcox, an African American songwriter, spent one growing season in the Bootheel.[18] After he left, Owen Whitfield, a Baptist minister and sharecropper, continued organizing farmworkers in the area. Despite local taboos against interracial friendships, Whitfield and Snow became colleagues.[19]

16. According to Pressly and Scofield, *Farm Real Estate Values,* the average value in dollars per acre of land in Mississippi County was $64 in 1910, $132 in 1920, $91 in 1925, and $37 in 1935. Records on file in the office of the Recorder of Deeds of Mississippi County, Missouri, indicate that Snow defaulted on several loans, lost property to foreclosure, and sought help from his wife's brother (see Chapter 2).

17. Thad Snow to Julien Friant, November 4, 1934, Julien N. Friant Papers, Regional History Collection, Southeast Missouri State University, Cape Girardeau.

18. Rebecca B. Schroeder and Donald M. Lance, "John L. Handcox, 'The Sharecropper Troubadour.'"

19. Delaney interview, August 16, 1999. Mrs. Delaney reported that Whitfield was a frequent visitor at Snow's Corner and that Snow often visited Whitfield after he settled at Cropperville in Butler County.

Whitfield, an African American man with a slender physique, a gravelly voice, and a powerful presence,[20] led the sharecroppers in the famous roadside demonstrations in the Bootheel in 1939. In mid-January, black and white farmworkers and their families pitched tents and set up camps along U.S. Highway 60 between Sikeston and Charleston and U.S. Highway 61 between Sikeston and Hayti in protest against the eviction of tenants from the land. Local planters reacted with outrage. The Missouri State Highway Patrol dispersed the demonstrators, who sought refuge in the towns, cities, and waste places of the Bootheel.[21] Whitfield fled to St. Louis and found supporters in that city.

The demonstrations raised public awareness of the share-croppers' troubles. The St. Louis Committee for the Rehabilitation of the Sharecroppers funded an experimental community, called Cropperville, as a refuge for displaced farmworkers, west of Harviell in Butler County.[22] The federal government constructed several housing projects in the Bootheel, and some of these remained viable communities throughout the twentieth century.[23] But displacement of tenant farmers continued as planters turned to day labor and then replaced farmworkers with increasingly efficient farm machinery.

Snow did not actively participate in planning or carrying out the roadside demonstrations. Whitfield's daughter, Shirley Whitfield Farmer, remembered Snow as a presence on the periphery. As a young girl, she occasionally saw him speaking with her father and

20. Dan Cotner, interview with the author, Cape Girardeau, Mo., April 4, 2002; Dr. Cotner was Owen Whitfield's dentist in the 1950s and 1960s. A small man himself, Cotner remembered Whitfield as being very slight, almost frail. Mr. Robinson remembered that Whitfield was a dynamic preacher but not a really good speaker, because he "talked with a scratch in his throat" (Robinson interview, September 10, 2002).

21. For a book-length study of the sharecroppers' roadside demonstrations, see Louis Cantor, *A Prologue to the Protest Movement: The Missouri Sharecropper Roadside Demonstration of 1939*. See also Lorenzo J. Greene, "Lincoln University's Involvement with the Sharecropper Demonstration in Southeast Missouri, 1939–1940"; and Arvarh E. Strickland, "The Plight of the People in the Sharecroppers' Demonstration in Southeast Missouri."

22. See Jean Douglas Cadle, "Cropperville, from Refuge to Community: A Study of Missouri Sharecroppers Who Found an Alternative to the Share-cropper System."

23. See Steve Mitchell, "Homeless, Homeless Are We . . ."

some other "old men" in their front yard at Cropperville.[24] For most of the year preceding the demonstrations, Snow was traveling in the West and in Mexico. But during the roadside strike, rumors spread that Snow had somehow "masterminded" the event.

He responded to the charge with a sarcastic "True Confession" in the *St. Louis Post-Dispatch,* admitting that he and Leon Trotsky and host of others, including his young daughter Emily, had plotted together in a vast conspiracy.[25] His daughter Fannie insisted in an interview that he actually had met and talked with Trotsky in Mexico, while she and Emily waited outside in the car.[26] Whether or not he actually spoke with the Russian revolutionary, Snow did not play a central role in the roadside strike. In spite of his strident prose, he remained a member of the privileged class. He protested by writing to his friends in Washington rather than by demonstrating on the highways. Although he respected the black strikers and their leaders, he spoke of them in terms that revealed a deep-seated paternalism.

Snow's experience with the sharecroppers' insurgency led him to examine his own, and his nation's, assumptions about race, justice, and war. He crossed a line in the late 1930s when he supported the interests of farm laborers against those of his own class. Events of the 1940s challenged his beliefs and drew him further from the mainstream of American political orthodoxy. He recognized the absurdities of fascism and condemned Hitler's anti-Semitism in no uncertain terms. But he also questioned the core assumptions of American liberal capitalism. He believed that commercial interests ruled the world, bringing misery and ruin to the farmers and the poor. War, in his new philosophy, was the inevitable result of economic competition among nations. He opposed the United States' involvement in World War II and, later, in the Korean War. But he clung to the hope that ultimately human beings would discover a way to resolve conflicts without resorting to war.

Tragically, he could not protect his own family from violence. In a bitter suicide note, his son-in-law Hartwell Thompson blamed Snow for the deaths of Priscilla, Ann, and Emily, saying "Thad Snow, you

24. Shirley Whitfield Farmer, interview with the author, Jackson, Mo., April 7, 2002.
25. Snow, *From Missouri,* 262–64.
26. Delaney interview, August 16, 1999.

will get some of the torture you have caused us."[27] In placing himself in opposition to the opinions of many of his neighbors, Snow may also have alienated his son-in-law, unwittingly helping to cause the deaths of his daughters and granddaughter.

At the end of his life, he found a new frontier in the Missouri Ozarks, where American backwoods culture endured throughout the twentieth century. When Fannie and her husband came home to Snow's Corner and took over the farm, Snow began spending most of his time away from the Bootheel in his room at the Rose Cliff Hotel, where he could look out his window at the deep wooded valley of the Current River.

When he died in 1955, he left an important legacy. In dozens of short articles for the *St. Louis Post-Dispatch,* he recorded the experiences of Bootheel farmers and croppers during the Great Depression. He stood out among his neighbors as a planter who sympathized with farm laborers, both white and black. Handcox, a labor poet, memorialized him in a song. Snow's book, *From Missouri,* was published in 1954, a few months before his death. In this book, which represented fifteen years of effort by a man in declining health, he attempted to tell not only his own story but also the story of the region he called "Swampeast Missouri."

Snow examined his own life, and in so doing took a critical look at some cherished values. When he first came to Missouri, he pursued the American dream in its most archetypal form as a twentieth-century frontiersman, seeking wealth by making a fresh start on a piece of undeveloped land. He died wondering if, in taming the frontier, Americans lost more than they gained. He also wondered—although he clung to a stubborn optimism—if America was capable of making one more fresh start.

---

27. *St. Louis Post-Dispatch,* August 15, 1948; *Sikeston (Mo.) Standard,* August 16, 1948; *Cape Girardeau Southeast Missourian,* August 16, 1948.

Snow's Corner   1

Twelve years ago a young man pulled up stakes in Indiana and headed for Mississippi County, Missouri. . . . [A]nd if you should ever travel from Bird's Point, opposite Cairo, Ill., to Charleston and then on toward Sikeston to the Mississippi county line you'd be on a concrete road obtained largely through his efforts. The same is true of the other cross-county road now in use. Considering this, . . . and his community work in general, we feel that Indiana once lost a mighty valuable man, and that now Missouri has a better one.

—A. I. Foard [1923]

Snow's farmhouse stood at an intersection in a region where roads stretched flat and vanished at the river or the horizon, and fence lines were section lines running straight as grids on a map. His farm occupied a tract of this flat land, flood land, alluvial plain, washed, drenched, and replenished by overflowing rivers and streams. Snow called it delta land, because it was formed, like the southern delta, from soil cast off by the Mississippi River. Before 1910 it was a wetland, thickly forested with cypress, oak, and other hardwoods.

Snow came to this land because it offered him the challenges and opportunities of an agricultural frontier. He left no great mark on Hancock County, Indiana, where he farmed for nearly a decade. In 1910, he served as president of the progressive Farmer's Institute there. Because he left the county in that year, another man had to complete his term.[1] His neighbors thought enough of him to choose him as a leader, but he was still a young man when he left Indiana, looking for success, personal distinction, and adventure.

He grew to love the Bootheel, and he became part of it. It would be impossible to understand his life without comprehending the peculiar culture of the southeastern Missouri lowlands. Conversely, a study of his life reveals much about society in this troubled region at a time when it stood at a crossroads—between forested wilderness and cut-over fields, between midwestern diversified agriculture and southern cotton farming, between backwoods traditionalism and an uneasy transition to modern American life.

When Snow arrived in the Bootheel in 1910, an amateur archaeologist named Thomas Beckwith was collecting and cataloging projectile points, tools, pots, and effigies from the Indian mounds on his family farm near Charleston. Shortly before his death in 1913, Beckwith donated his outstanding collection to Southeast Missouri State University in Cape Girardeau. The artifacts in that university's museum collection document the Mississippian civilization that flourished between A.D. 900 and A.D. 1700 in the lowlands. The Mississippians depended on the river for fish and on the rich bottomlands for raising corn, beans, and squash. A wide-flung trading network connected their villages with the cultural center at the confluence of the Missouri and Mississippi Rivers. For reasons that remain mysterious, their civilization declined before European settlers appeared in the area, but they left their mounds and artifacts as reminders of the past.[2]

After the American Revolution, a New Jersey veteran named Colonel George Morgan dreamed of building a great commercial center below the mouth of the Ohio River in Spanish territory on the western bank of the Mississippi. He obtained a land grant from

1. On January 12, 1998, Carolyn Autry, a reference librarian for the Indiana Historical Society, wrote to the author, saying, "I regret that I have found virtually nothing relating to Mr. Snow. The only mention of him was in George J. Richman's *History of Hancock County, Indiana*, on page 172, where it was noted that he was President of the Farmer's Institute in 1910. Apparently he did not serve for the entire year since another man was mentioned as the president in 1910."

2. "Beckwith Collection: An Illustrated Vignette of the Mound Builders, with Photographs"; Carl H. Chapman and Eleanor F. Chapman, *Indians and Archaeology of Missouri*, 71–89.

Spain and planned the new city on a grand scale with wide streets and magnificent plazas. With the help of a small group of settlers, he established the town in 1789. Jealous rivals maliciously set out to destroy him, and within a year the Spanish government stripped him of his land concession and sent him packing.[3]

Spanish soldiers soon constructed a fort at Morgan's town of New Madrid, which became the center of government for a large district on the west bank of the Mississippi. Nicolas de Finiels, a French mapmaker working for the Spanish monarchy, described the terrain between New Madrid and Cape Girardeau as a "chaos of trees, water, and mire."[4] At the confluence of the Mississippi and Ohio Rivers, Finiels encountered a fertile land, submerged at high water, but dry in the summer months. He found settlement sparse, even on ridgetops, for lack of a reliable water supply.

During the early nineteenth century, settlers continued bypassing the southeastern "swamp country" because, in the words of one early inhabitant, it had a reputation for being "sickly and visited by earthquakes." The New Madrid earthquake of 1811 certainly damaged the terrain, shifting the courses of rivers and streams, creating lakes, ponds, and sinkholes. But the major barrier to settlement remained the periodic inundation of wide stretches of the region when the Mississippi and the St. Francis Rivers and other smaller streams flooded.[5]

By the 1850s, American entrepreneurs regarded swamplands as opportunities for investment and development. Under the Swamp Land Acts of 1849, 1850, and 1860, the United States Congress turned over millions of acres of wetlands to state governments. Many states subsequently formed drainage districts, which were public corporations that had the power to determine land ownership, claim eminent domain, issue bonds, construct ditches and levees, and recover costs by taxing landowners. Large landowners exercised the most influence and derived the greatest benefit from reclamation

---

3. *Goodspeed's History of Southeast Missouri* (1888), 284–86.
4. Nicolas de Finiels, *An Account of Upper Louisiana*, 33.
5. "Judge Goah Watson's Account of the Settlement of New Madrid," typescript, collection 995, vol. 3, no. 1001, pp. 11–13, Western Historical Manuscript Collection, University of Missouri–Columbia; James Lal Penick, *The New Madrid Earthquakes*, 89; Leon Parker Ogilvie, "Governmental Efforts at Reclamation in the Southeast Missouri Lowlands," 151, 159.

projects. Due mostly to the work of these drainage districts, between 1850 and 1930, the Midwest lost 70 percent of its wetlands.[6]

Before the Civil War, regional boosters urged Missouri's state government to support swamp drainage. In 1851 Representative Robert
A. Hatcher of New Madrid County pushed a bill through the legislature establishing a special reclamation district composed of counties in southeastern Missouri containing swamps or overflowed lands.
Within two years, however, the state turned responsibility for the
drainage program over to the governments of six Bootheel counties: Dunklin, Mississippi, New Madrid, Pemiscot, Scott, and Stoddard. In explaining this decision, Governor Austin A. King cited a
lack of cooperation from county officials.[7]

While Bootheel residents struggled with the drainage problem,
loggers cleared millions of acres of timberland in other parts of the
United States. The westward movement, industrialization, and urbanization created an unprecedented demand for lumber. Beginning in the 1850s, steam-powered circular and gang saws increased
the output of lumber mills from less than three thousand to more
than forty thousand board feet per day. American loggers cut less
than two billion board feet of wood in 1839, more than eight billion
in 1859, twenty billion in 1880, and a peak of forty-six billion board
feet in 1904, an amount never again equaled.[8]

The Civil War slowed economic development in the Bootheel.
Many of the local inhabitants came from or could trace their ances-
tors to Virginia, Kentucky, Tennessee, and other slave states. Early
in the war, Confederate forces occupied New Madrid. Partisan activity and guerilla warfare troubled the region. Mississippi County
leaned strongly toward the Confederacy. Federals attacked the
Southern encampment at Belmont, on the river near Wolf Island, in
the fall of 1861, but the Confederates drove them away.[9]

After the war, Louis Houck opened up the Bootheel by investing

6. Donald J. Pisani, "Beyond the Hundredth Meridian: Nationalizing the
History of Water in the United States," 476.

7. Ogilvie, "Governmental Efforts," 165–76.

8. Michael Williams, "The Clearing of the Forests," in Michael P. Conzen, ed.,
The Making of the American Landscape, 152–53.

9. Eileen Meinershagen, "Concerning the Early History of Mississippi County"
(n.p., 1952), typescript on file at the Mississippi County Public Library;
Goodspeed's History of Southeast Missouri, 498. The Battle of Belmont was indeci-
sive, although the human cost was high. On November 7, federal troops
crossed the Mississippi River from Cairo and drove Confederates out of their

in short-line railroads connecting its towns and forests to Cape Girardeau, St. Louis, and the wider world. The Illinois-born lawyer and entrepreneur arrived in southern Missouri in 1869. In the spring of that year, he traveled south from the site of Dexter in Stoddard County to Kennett in Dunklin County and marveled at vast stretches of primeval forest. Late in his life he recalled his impression:

> The Kennett road passed down [Duck] creek and I remember before I went into the bottom, I sat down on a hillside on the left of where Dexter now stands, looking over a vast forest of timber on all sides, greatly impressed; not a single farm in sight or opening in the vast woods except at the foot of the hill, an open place known as Miller's Farm. Going south for miles we traveled along the edge of what was known as the East Swamp, all covered with heavy timber.[10]

Houck spent the next three decades of his life acquiring property, promoting development of the region, building roads and railroads, and encouraging lumber companies to harvest the timber.

Houck's railroads opened the lowlands to the lumber industry. The short lines he constructed in the 1880s and 1890s connected the wooded regions of the Bootheel with timber markets in Cape Girardeau and elsewhere. His correspondence contains numerous exchanges with lumber companies considering investing in the area. Many timber enterprises, including Gideon and Anderson, Himmelberger-Luce, C. A. Boynton, and International Harvester, built sawmills and planing mills along Houck's lines. Lumber companies acquired huge tracts in southeastern Missouri. For instance, Himmelberger and Harrison of the Luce Land and Lumber Company controlled two hundred thousand acres of Bootheel land.[11]

---

earthen fortifications, but the Southerners quickly rallied and forced the enemy to flee. Both sides claimed victory, and both sides suffered heavy casualties (610 Union and 641 Confederate soldiers killed, wounded, captured, or missing, out of a total of 3,000 Union and 5,000 Confederate troops). See "Battle of Belmont," *Charleston (Mo.) Enterprise-Courier,* special publication, October 1991; and *Goodspeed's History of Southeast Missouri,* 501–5.

10. *Cape Girardeau Southeast Missourian,* May 15, 1969; newspaper edition of Houck's reminiscences, bound in manuscript, in possession of the Houck family.

11. William T. Doherty Jr., *Louis Houck: Missouri Historian and Entrepreneur,* 43–44; Bonnie Stepenoff, "The Last Tree Cut Down: The End of the Bootheel Frontier," 64; Louis Houck Papers, Regional History Collection, Southeast Missouri State University, Cape Girardeau.

Lumber companies brought new workers into the Bootheel forests. One of the largest, the Wisconsin Lumber Company, a subsidiary of International Harvester, began operating in Pemiscot County around 1902. Workers migrated from east of the Mississippi River to work in the company's sawmill in Deering. The company owned most of the town; workers rented housing from the company and shopped at the company store. According to oral testimony, the company hired both black and white workers. Although black workers traded at the company store, they lived in a separate area called Negro Town, or Colored Town, with their own school, church, hotel, and recreation center. Deering's post office was located in the company store. In addition to the store, the company provided a barbershop, an icehouse, a ball diamond, and a playground.[12]

Like other lumber milling towns of the early twentieth century, Deering occupied an isolated clearing in the wilderness. Early resident Mary Putnam Williams remembered that the roads were so bad before the 1920s that it might take two hours to travel two miles. The marshy ground stuck to the wheels of a wagon, impeding progress. After a rain, the roads remained muddy until the sun baked them dry; shade from overhanging tree limbs made this a slow process. Wild animals abounded. When the mill whistle blew early in the morning, wolves would howl in response. People had to fence in their livestock to protect it from bobcats and other predators.[13]

Lumber-mill workers adopted the hunting and gathering lifestyle of the indigenous population. As Williams recalled, people could fish in a big ditch south of Deering. She remembered that black women often fished along the banks. Apparently, white families also depended upon fish for many meals, and in the winter people killed wild turkeys and hogs for meat. "The wild hog meat wasn't as good as the fattened hogs," Williams said, "But it sure beat nothing."[14]

Ironically, the success of lumbering and related industries, such as the box factory at Gideon, stripped the wilderness of game and the people of subsistence. As lumber towns boomed and died, most of the timber workers found work as farm laborers, headed for fac-

12. Ophelia R. Wade, ed., *History of Delta C-7 School District in Deering, Missouri* (Deering, Mo., 1976), 51–52.

13. Quoted in ibid., 52.

14. Ibid.

tories in cities and towns, or moved on in search of an elusive frontier. Only a lucky few succeeded in acquiring land and becoming farmers.[15]

By 1905, the Bootheel's lumber boom propelled a massive engineering feat that would alter the landscape forever. In January of that year, supporters of economic development formed the Little River Drainage District, a public agency that planned and constructed a system of ditches, canals, and levees to drain the swamps. District boundaries stretched from Cape Girardeau southward to the Missouri-Arkansas state line, covering an area of some 540,000 acres in Bollinger, Cape Girardeau, Dunklin, New Madrid, Pemiscot, Scott, and Stoddard Counties. Mississippi County formed its own drainage district. Property owners paid taxes to support facilities that eventually drained or provided drainage outlets for more than a million acres of swampland.[16]

As drainage proceeded, farmers seized the opportunity to acquire large tracts of rich, uncultivated land in the Bootheel. For example, Xenophon Caverno, a Wisconsin businessman, established Headlight Plantation at Canalou (New Madrid County), Missouri, about 1907. An 1890 graduate in mechanical engineering from the University of Wisconsin, he had a varied career as a machinist and draftsman for a railroad, the owner and manager of a daily newspaper, and a manufacturer of equipment used to provide utilities for farm homes. Finally, he became a farmer and joined what he called "a wave of immigration from the north attracted by the rich black soil" of the Bootheel.[17]

Thad Snow joined this wave of immigration a few years later. Like Caverno, he was a northerner who felt the attraction of the rich soil of the Mississippi River bottoms. The Bootheel offered a longer growing season than Hancock County, Indiana, so it was possible to produce bigger yields of traditional crops. Snow became one of the first farmers to grow alfalfa in Mississippi County, where it was possible to harvest three crops off the same acreage in the same season. Later, he would join other southeast Missouri farmers in

15. Stepenoff, "Last Tree," 71.

16. *Little River Drainage District of Southeast Missouri 1907–Present*, 6.

17. Xenophon Caverno to Robert Jordan, February 8, 1939, folder 51, Xenophon Caverno Papers, Western Historical Manuscript Collection, University of Missouri–Columbia.

turning their acreage to cotton production. In addition to experi-
menting with new crops, Snow pushed for highway construction
and economic growth in the Bootheel.[18]

Land acquisition, drainage, and crop development involved Snow
in a complicated series of mortgages and legal transactions. In April
1911, he and his financial partner, Anna D. Cooper, an Indiana
widow, successfully sued Humphrey C. W. Stanley and other members of the Stanley family to confirm a clear title to a tract of just
under eleven hundred acres.[19] The Stanleys claimed ownership of
the land on the basis of Swamp Land Patents dating from the 1860s,
but the court ruled against them.[20] About a year later, Snow and
Cooper mortgaged the tract for $30,000, signing a deed of trust to
the Prudential Insurance Company of America.[21] After paying off
the first mortgage within five years, Snow and his partner obtained
another mortgage of $38,000 from Phoenix Mutual Life Insurance
Company.[22] Cooper apparently was merely an investor, not an active partner in the farm. In 1926, Snow was finally able to buy out
Cooper's interest in the property for $20,000.[23]

As a landowner with a thousand-acre farm, he belonged to an
aristocratic group in the Bootheel, although he was far from the
largest landowner in the region. One of his close friends was Xenophon Caverno, who owned two thousand acres in a neighboring
county.[24] Both Caverno and Snow were associates of Charles D.
Matthews, who became president of the Bank of Sikeston in 1911. A
son of a large landowner, Matthews was also president of the Baker-
Matthews Lumber Company and vice-president of the Scott County
Milling Company.[25] In the early 1930s, Matthews, Caverno, and Snow

18. A. I. Foard, "He's Putting Alfalfa thru Its Paces," undated clipping in
Giboney Houck, Incoming Letters, folder 1923, Louis Houck Papers.
19. Mississippi County, Mo., deed book 64, p. 342. Cooper's name appears
on deeds recorded in Mississippi County, but information about her is scarce.
A Mrs. D. B. Cooper is listed in the 1920 telephone directory for Greenfield,
Indiana, which is on file at the Hancock County Public Library in Greenfield.
20. Circuit Court Record, Mississippi County, Mo., book 22, p. 562, February
9, 1911.
21. Mississippi County, Mo., deed book 59, p. 60.
22. Mississippi County, Mo., deed book 49, p. 476, and deed book 98, p. 505.
23. Mississippi County, Mo., deed book 98, p. 505.
24. Caverno Papers, register.
25. State Highway Commission of Missouri, *Sixth Biennial Report* (Jefferson
City, Mo., 1928), [2].

jointly headed a "group of representative farmers of Southeast Missouri," who contacted the U.S. Department of Agriculture concerning problems with federal policy.[26]

Snow and Caverno shared similar backgrounds but grew farther and farther apart in their political views. Describing his upbringing, Caverno wrote, "I was born in Wisconsin into the traditions of New England abolitionism, the memories of the Civil War, and the religion of Republicanism."[27] Caverno upheld these traditions throughout his life. Snow, too, was reared in the tradition of Civil War–era Republicanism. Before migrating to southeastern Missouri, however, Snow abandoned the Grand Old Party, generally favoring the Democrats, but gradually evolving as an independent political thinker, who flirted with the Progressive Party, made friends with Socialists, and supported and then criticized the Democrats' New Deal. In the late 1930s, Snow's political views became more radical, resulting in friction with Caverno, but the two men and their families maintained their friendship.[28]

Throughout his life in the Bootheel, Snow remained a family man, connected to kin in Indiana. Tragically, his first wife, Bessie, died of a hemorrhage after a fall from a horse in 1915, when their son, Hal, was ten and daughter, Priscilla, was eight years old. After her death, Bessie's sister, Alma, and brother-in-law, Lawrence Catt, moved from Indiana to a neighboring house at Snow's Corner to help care for the children.[29] Aunt Alma remained in Mississippi County, raised her own family, and had a close relationship with Snow's children and grandchildren by both his first and second marriages.[30]

26. X. Caverno, C. D. Matthews, and Thad Snow to Hon. Julien N. Friant, Telegram, November 23, 1934, Caverno Papers, folder 59.

27. Caverno to Herbert Croley, November 16, 1927, Caverno Papers, folder 64.

28. Delaney interview, August 16, 1999. In the 1920s, people still assumed Snow was a Republican. Albert Ross Hill, president of the University of Missouri, referred to Snow as "an up-to-date, first-class Republican," until Snow corrected him. Hill to Snow, December 14, 1920, University of Missouri President's Office Papers, Western Historical Manuscript Collection, University of Missouri–Columbia, folder 1331.

29. *History and Families, Mississippi County, Missouri, 1845–1995*, 123–24; Nunnelee Funeral Home (Death Records), Mississippi County, Missouri, 1910–1930; United States Census, Mississippi County, Missouri, 1920.

30. Debbie Delaney Corse, telephone interview with the author, May 28, 2002.

In 1919 Snow's parents, Henry and Sarah Frances ("Fannie"), came to the Bootheel and purchased a house in Charleston.[31] Two years later, Henry died, leaving all his property to his wife. In his will, Henry noted that his only surviving sons were Ralph Snow of California and Thad Snow of Missouri. "I give nothing to either of them, they being well able to take care of themselves."[32] Thad Snow was executor of the will. His widowed mother at first remained in Charleston but by 1930 had moved to the farmhouse at Snow's Corner. According to her granddaughter and namesake, Fannie was a well-read woman who introduced Snow's children to classics such as Dante's *Inferno*.[33]

Snow met his second wife, Lila, through the poems of Elizabeth Barrett Browning. She had a copy of *Sonnets from the Portuguese;* he heard about it and asked if he could borrow it. In the later years of their marriage, she became a sounding board for his evolving political philosophy.[34] Their elder daughter, Fannie, vividly remembered her parents reading and discussing books: "[They] spent their time with their noses in books when he wasn't farming."[35]

Snow loved farming, and he developed a particular fondness for trading, breaking, and training mules. Late in his life, he credited mules with human virtues, especially the capacity for hard and dedicated labor. According to Snow, this was one of mankind's most admirable qualities, which was shared with a four-legged cross between a donkey and a horse. Of his favorite mule, Snow wrote nostalgically, and only slightly sardonically, "Kate was admired and bragged about by every man who worked her or ever saw her work. It didn't turn her head. She mellowed a little, I think, as the years went by; but she remained aloof to the end. Her work was her interest in life, and she gave it all she had."[36] He might have applied these words to his own personality.

31. Mississippi County, Mo., deed book 79, p. 439.

32. Henry Snow, Last Will and Testament, Mississippi County, Mo., deed book 94, p. 44.

33. The 1930 census lists the members of Snow's household as Snow, 48, Lila, 42, Lena Frances, 8, Emily, 3, and Frances Pierson, 81. U.S. Bureau of the Census, *Fifteenth Census of the United States: 1930, Population Schedule: Mississippi County, Missouri.* Delaney interview, August 16, 1999.

34. *St. Louis Post-Dispatch,* May 2, 1937.

35. Delaney interview, August 16, 1999.

36. Thad Snow, "Proud Kate, the Aristocratic Mule," 68.

In his early Bootheel days, he took great pleasure in mingling with other men in the big sales barns where farmers bought mules. As Snow recalled, when he and his neighbors were clearing the land using axes and mule power, mule dealers were the biggest businessmen in Bootheel towns. They shipped in young unbroken mules from the north and shipped the old, worn-out animals farther south to the cotton lands. Pioneer farmers bought them on credit, broke them in, and sometimes lost them for nonpayment of their debt.[37]

About 1913, Snow hired Mose Feezor, a white farmer from Kentucky, to help him cut trees, manage his livestock, and supervise the farmhands. Hardworking and stubborn, Feezor was an excellent mule handler. Snow remembered one log that measured five and a half feet in diameter. Feezor managed to load it on a wagon with two mules, chains, and long wooden skids. Feezor worked at Snow's Corner for more than twenty years. His daughter, Nellie Feezor Stallings, recalled many years later that he tended the cattle, hogs, and mules and made sure the other farmhands were working on time and where they were supposed to be. He never went out into the fields to work, however; that was the job of other hired hands.[38]

Stallings remembered Snow as an odd, aloof, but kindly man, who functioned more like a plantation master than an employer. He provided a house for Mose Feezor and his wife, Sarah, and their four daughters and two sons. Both boys were born at Snow's Corner. Although Stallings said Snow was "always nice to us kids," he kept his distance. "Of course," she said, "we weren't around him a lot." In a paternalistic fashion, however, he came to pick up the four girls when their first brother was born, took them to the big farmhouse, and told them, "Now you stay here. Your mother's sick, and I'm going down to see about her."

He worried about the children's safety. The Feezors attended Grigsby School, a one-room country school about four miles from Snow's Corner. If it was raining or snowing, Snow would go to school and pick them up in the wagon. "And then," recalled Stallings, "a lot of times he would ride his horse that he always rode,

---

37. Ibid., 65.
38. Nellie Feezor Stallings, interview with the author, Charleston, Missouri, April 1, 2001. Quotations in subsequent paragraphs refer to this interview.

and he'd bring two old mules for me and my three sisters to ride."
When the state was building the highway from Charleston to Cairo,
Snow would "meet us and walk us by all the men, because he didn't
think it was safe. And of course Mama couldn't come after us, because that wouldn't have been good either."

The Feezor children had limited contact with Snow's family.
According to Stallings, the Snow children went to school in Charles-
ton, while the Feezors attended the rural school. Local records con-
firm that all of the Snow offspring graduated from Charleston High
School (Hal in 1922, Priscilla in 1925, Lena Frances in 1939, and
Emily in 1943).[39]

Stallings was clear about class divisions at Snow's Corner. Snow
talked to her father in a friendly way, but they were not equals, and
the children knew it. As she later recalled, "Well, he talked to us
kids a right smart, but you know, we were always kind of a little
afraid of him. He was always nice. He never said anything out of the
way. But we knew to stay away from him. We didn't fool with him."

She remembered when Snow started raising cotton on his farm,
although she didn't recall what year it was. That was when African
American families came to live at Snow's Corner: "I remember the
colored people coming in and raising cotton. But we didn't, you
know, the kids didn't fool around with them." At a very young age,
she knew where the social and racial boundaries were located.

From her vantage point at Snow's Corner, Stallings witnessed a
great migration as white and black lumber workers and farm laborers streamed into the Bootheel, hoping for a better life. In a time of
urbanization, while Missouri's rural population declined, the population of the Bootheel swelled.[40] In many cases, the instigators and
beneficiaries of this population boom were lumbermen, real estate
promoters, land speculators, and absentee landlords, who had no
long-term commitment to the region. "Swampeast Missouri" became a region of wealthy landowners, struggling tenants, poverty,
transiency, and racial conflict.[41]

39. "Graduating Classes of Charleston High School, 1906–," typescript on file
at the Mississippi County Public Library, Charleston, Mo.

40. U.S. Bureau of the Census, *Thirteenth Census of the United States: 1910*, vol.
2: *Population*, 1106, 1114; *Fifteenth Census of the United States: 1930*, vol. 3, pt. 1:
Population, 1341–46.

41. David Thelen, *Paths of Resistance: Tradition and Democracy in Industrializing
Missouri*, 92.

Crime, disorder, and vigilantism increased as poor white and black workers migrated to the region seeking employment in the lumber mills and box factories and on the farms. On July 8, 1910, the year Snow came to the Bootheel, the *Charleston Enterprise-Courier* reported a double lynching on the preceding Sunday afternoon. The victims were two African American men who allegedly robbed and murdered William Fox, a wealthy farmer in the county. A mob stormed the jail and seized and hanged the two men, while, according to the newspaper, "Almost the entire populace was out, as were farmers for miles around, to witness the affair."[42] In Caruthersville (Pemiscot County), a mob murdered a black man named A. B. Richardson and burned a black boardinghouse in the fall of 1911. By 1915 large landowners had banded together to force local authorities to prosecute white vigilantes and protect the biracial labor force.[43]

The process of logging off and clearing the land created a restless and exploited work force. In his early years in Mississippi County, Snow hired a succession of men and families to do the backbreaking work of cutting trees and hauling out logs. Through the prism of his own experience, he viewed them as hardworking pioneers, but they were merely laborers with little hope of bettering their position. They were paid by the acre and not by the day. The more land they cleared, the more they earned. Yet, a terrible irony existed: Cutting trees off a piece of land meant cutting off their means of livelihood. As the trees disappeared, laborers moved away or tried to find work on local farms.

Having exhausted huge tracts of timber, lumber companies attempted either to join in the agricultural economy or to sell their holdings to large planters. Charles B. Baker recalled that in 1928 the International Harvester's Deering mill went idle. Subsequently, the company tried to turn its large landholding into a family farming operation, using former mill personnel as sharecroppers. Failing dismally, the company sold over twenty-five hundred acres of cleared land to Baker in 1935. Baker employed tenant farmers to raise cotton, soybeans, and alfalfa.[44]

Tenant farming endured as a way of life in the Bootheel. After

42. *Charleston (Mo.) Enterprise-Courier,* July 8, 1910.
43. Thelen, *Paths of Resistance,* 96–97, 99.
44. Interview with Charles B. Baker, April 24, 1976.

twenty years with Snow, Mose Feezor went into farming for himself. But he did not own his land; he rented it. His daughter Nellie married Marshall Stallings, who rented and farmed acreage at Concord and Gravel Ridge. Social distinctions between landowners and tenant farmers remained strong. In her adult life, Nellie Stallings had very little contact with Snow. Her husband, Marshall, may have had some business dealings with Snow. There was never any trouble between them. But, as she explained, "Well, really, he [Marshall] was a working man, and Snow was [pause] something else. I'm sure that made a difference."[45]

She saw Snow not as a member of the working class but as part of a privileged elite. Although he viewed himself as a pioneer, he began with a large investment and, despite mortgages and setbacks, he watched his investment grow. He was a compassionate person, and he sympathized with the laborers on his farm. In the 1930s, he risked his reputation and his friendships by championing their cause, but in the end they were his workers, and he was, in Stallings' words, "something else."

Nearly half a century after his death, the farm at Snow's Corner continued to prosper. Snow's granddaughter and her husband farmed it by themselves with the aid of modern machinery. They raised soybeans, not cotton, and preserved several tracts of woods.[46] Snow's daughter Fannie lived in a house that she and her husband built in the 1950s, just forty feet from the house in which she was born. She did not share her father's urge to move to a new and undeveloped land.

The essayist Wendell Berry, echoing historian Wallace Stegner, identified two distinct types of human beings: "boomers" and "stickers." Snow, at least in his early years, was a boomer, who wanted to cash in on the resources of a new land. His daughter, perhaps absorbing lessons her father learned, became a sticker, who valued the land on its own terms. In her farmhouse at Snow's Corner, in 1999, Fannie Snow Delaney said, "My roots here are so deep, I can't even imagine going anywhere else."[47]

45. Stallings interview, April 1, 2001.
46. Corse interview, May 28, 2002.
47. Wendell Berry, *Life Is a Miracle: An Essay against Modern Superstition*, 131; Delaney interview, August 16, 1999.

The Big-Eye   2

Yet in the fall of 1910 I suddenly made a decision—I was going somewhere to a different sort of country. If I could find it I wanted it to be a
"new" country that I could spread out in, and one where a lot remained
to be done.

—Thad Snow, *From Missouri*, 90

Xenophon Caverno described the earliest settlers of the Bootheel
as followers of Daniel Boone.[1] He meant this not literally but figura-
tively. Throughout American history, many hardy followers of
Boone have ventured into waste places, mountainous places, arid
places, swampy places, lonely places, searching for a piece of land
on which to make a home, or a name for themselves, or a fortune.
Others have simply succumbed to a wanderlust, seeking enlighten-
ment, adventure, or novelty, or not really knowing what they were
looking for at all. Caverno and other twentieth-century migrants to
the Bootheel clearly hoped to cash in on the rich soil being uncovered by logging operations.

In 1955, the local newspaper eulogized Snow as a "rugged individualist," motivated in the early twentieth century by the "pioneer
urge."[2] This phrase, of course, romanticized and oversimplified
Snow's decision to move to the Bootheel, but there was some truth

1. Xenophon Caverno to Robert Jordan, February 8, 1939, Caverno Papers,
folder 59.
2. *History and Families, Mississippi County*, 143–44.

to it. Like the men of Boone's era, Snow traded a settled existence for a chance to remake a tract of forested land and turn it into a prosperous farm. He pulled up stakes, moving his wife and children. Eventually other members of his family, including his parents, left Indiana and joined him in Missouri.

In part, he may have been seeking freedom and a closer relationship to nature. Throughout his life, he was an avid horseman and hunter. In 1941, he wrote in an editorial for the *St. Louis Post-Dispatch* that he had just come back from a four-month "jag"—not a drinking jag, a shooting jag: dove shooting, duck shooting, quail shooting. He found his own behavior puzzling and wasteful, but "intoxicating." He wondered why civilized people tolerated it, and then he argued that

> Probably their tolerance derives from times in the dim past when a man was a good egg if he was a consistent game-getter, and not otherwise. Women like to dress and cook game. It makes them feel primitive, momentarily, and they like it. A man's self esteem touches the sky when he lugs a full bag after a good shooting day.[3]

This essay revealed a complex tangle of ideas connected to the Daniel Boone image but undercut by twentieth-century sophistication. Modern men and women merely "tolerated" such primitive behavior. Hunting allowed men to assert their masculinity; women enjoyed a temporary feeling of primitiveness when their men came home with game. But the assertions and feelings were momentary lapses among civilized beings. Snow's reference to the "dim past" indicated that he realized his behavior was anachronistic. The women he knew—his highly literate mother, for instance—hardly fit the hunter-gatherer mold of womankind. His own self-esteem surely did not depend on lugging home a dead animal's carcass. But he liked to keep the pioneer dream alive.

Novelists, historians, and social critics have generally agreed that the frontiersman was an archetypal American character, but they have disagreed on his essential qualities. Frederick Jackson Turner,

---

3. "Thad Snow Leaps from Jag to Jag," clipping in scrapbook, Thad Snow Papers, Missouri Historical Society, St. Louis.

lamenting the lost frontier, glorified him as the quintessential American, the progenitor of the individualism, pride, optimism, and resourcefulness without which democracy might not survive. Henry Chambers, writing about the settlers of the trans-Mississippi West, described them as "the inheritors of the wanderlust from those forebears whose ceaseless seeking out of frontier and mountain fastness from colonial times down had imparted to their descendants a restlessness which could be eased only by change."[4]

Recent writers have chided the frontiersman for his arrogance, his acquisitive nature, and his lack of a social consciousness. Wendell Berry, expressing environmentalist concerns, condemned him as a selfish exploiter of natural resources and, by extension, of his fellow human beings. Patricia Limerick and other new social historians of the West have questioned the frontiersman's identity—his whiteness and maleness—as well as his virtue.[5] These historians have portrayed the westward movement not as a romantic quest for freedom but as a great surge of wealth-seeking adventurers, emboldened by the imperialistic pride of an industrializing nation. Snow's character exemplifies this stereotype as much as it does the image of the innocent adventurer.

Looking back on his move to Missouri, Snow cast himself as an American pioneer in the Turner tradition. Certainly he defined himself as an individualist, in the sense of making up his own mind, sticking to his intellectual guns, and relying on his own resources. He was restless and ambitious, confessing to a desire to own a large piece of unbroken land, to develop it, and to make it pay. Like most human beings, he occupied a position between high idealism and cold calculation, altruism and undisguised greed.

He was an "inheritor of the wanderlust." His father, Henry, traveled and invested in enterprises in the West. Born in Kentucky in 1837, Henry was the son of English immigrants, Thomas P. Snow and Mary Ann Monks. At the age of nineteen, Henry Snow began teaching school in Hancock County, Indiana, where his parents had

4. See Frederick Jackson Turner, *The Frontier in American History;* and Henry E. Chambers, *Mississippi Valley Beginnings: An Outline of the Early History of the Earlier West,* 256.

5. See Wendell Berry, *The Unsettling of America: Culture and Agriculture;* and Patricia Nelson Limerick, *The Legacy of Conquest: The Unbroken Past of the American West.*

settled. He earned the title of captain as a volunteer in the Union army, Company F, Twenty-eighth Indiana, during the Civil War. Snow served in Missouri and Arkansas, taking part in the battle at Pea Ridge and suffering a wound at Petersburg during the Virginia campaign in 1864.[6]

After his discharge in 1865, Captain Snow returned to Greenfield, a rural village twenty miles east of Indianapolis, and taught for a year. Restlessness overtook him, and he went west to Warrensburg, Missouri, where he opened a jewelry store. Eleven years later, he returned to Greenfield, where he remained for the rest of his life, investing in various enterprises, including a cattle ranch in Indian Territory. He was a devoted Republican, a Mason of High Degree, a member of the G.A.R., and, while he was not active in the church, he was "a firm believer in revealed religion and in the efficacy of the church as a potent factor in civilizing and redeeming the human race."[7]

Captain Snow married Fannie Pierson, a Greenfield native, on October 3, 1877. Their son Thad was born on November 1, 1881. He had a sister, Lena, and a brother, Ralph. The siblings read Dickens, Scott, and Thackeray and raised a variety of pets, including chickens, pigeons, ponies, and dogs. Until Lena died in her teens, Snow enjoyed an idyllic childhood in a modest house on a large lot, with a ten-acre beech grove.[8]

Lena's death affected him deeply. He was thirteen, Lena sixteen, when she succumbed to typhoid fever. He named his second daughter Lena Frances, and he confessed in his memoirs that he never stopped thinking about his sister. He missed talking to her about books they had read. During the course of his life, he lost five women to untimely deaths, and he dedicated his book to them— Lena, Bess, Lila, Priscilla, and Emily. He particularly missed Lena, Lila, and Emily, the three women with whom he had shared his love of literature and ideas.

During his childhood and youth, Snow developed a deep appreciation for the natural world. He recalled late in life that, by the time

6. B. F. Bowen, *Biographical Memoirs of Hancock County, Indiana,* 351.
7. Ibid., 353.
8. Omer J. Walsh, *Boots and Walsh's Directory of the City of Greenfield, (1893– 1894),* 58, locates this house at 31 West Osage Street; Snow, *From Missouri,* 9.

he was ten, his father had made him acquainted with all the varieties of trees near his home. One of his father's friends was James
Whitcomb Riley (1849–1916), a Greenfield native who wrote elegiac
poems about nature, rural life, and the American backwoodsman.[9]

By the time Thad was born, Riley had left Greenfield and joined
the staff of the *Indianapolis Journal,* but he often came home and vis-
ited the Snows. In 1883, his first published volume of poetry, *"The
Old Swimmin'-hole" and 'Leven More Poems* appeared.[10] Although
Riley chose urban life for himself, he glorified farmers and pioneers.
For instance, in his poem, "A Tale of the Airly Days," he wrote:

> Tell me a tale of the timber-lands—
>   Of the old-time pioneers;
> Somepin' a pore man understands
>   With his feelins 's well as ears.
> Tell of the old log house,—about
>   The loft, and the puncheon flore—
> The old fi-er place, with the crane swung out,
>   And the latch-string through the door.[11]

Snow emulated the older man in several ways. Riley was a tire-
less conversationalist and storyteller,[12] and Snow, too, loved to visit
and entertain people with ideas and anecdotes. In his poems, Riley
created a cast of characters drawn from everyday life and all rungs
of the social ladder, but he had a special feeling for what Snow
called "lowly folks with quirks of one sort or another." These were
just the kind of people Snow liked to write about, people like the
door-to-door Bible salesman, who spent an hour in earnest conversation at Snow's Corner.[13] Both Snow and Riley had an affinity for
the kind of man exemplified by the poet's Jap Miller:

> Jap Miller down at Martinsville's the blamedest feller yit!
> When *he* starts in a-talkin' other folks is apt to quit!

9. Snow, *From Missouri,* 10, 24, 188.
10. *American National Biography* 18 (New York: Oxford University Press), 517-
18.
11. James Whitcomb Riley, *Riley Farm-Rhymes,* 155.
12. Peter Revell, *James Whitcomb Riley,* 21.

. . . .

Religen, law, er politics, prize-fightin' er baseball—
Jes' tetch Jap up a little and he'll post you 'bout 'em all.[14]

Families like the Snows and the Rileys created their own cultural
community in Greenfield. Singing, playing, and reciting poetry in
the parlor were common forms of entertainment. By the time Snow
was born, Greenfield had two literary societies, the Greenfield
Literary Club (founded in 1878) and the Greenfield Reading Club
(founded in 1879). Nurtured in this environment and watching his
hometown hero win national fame, Snow went to the University of
Michigan in 1902–1903 to study English composition. He did not
stay to complete his studies, allegedly because he did not want to
ruin his natural style of writing.[15]

Returning to Indiana, he married Bess and chose farming as his
occupation. The young couple set up housekeeping in a small cabin
on a farm owned by his father southeast of Greenfield. Within a
short time, Snow built a larger house and hired a farmhand to oc-
cupy the cabin. As his family grew, he developed a successful farming operation, raising hogs, sheep, corn, and alfalfa. He used mules
for plowing and dogs for herding sheep.[16]

In Indiana, in the early twentieth century, he farmed without the
benefit of machinery or electricity. On his own, or with a hired hand,
he milked three cows, took care of his horses, chopped hay, and
tended a flock of sheep. He fed his stock by lantern light. While his
dogs drove the sheep down the road to Greenfield, he waited and
loafed with his friends on Main Street.[17]

While living in Hancock County, Snow took an active role in the
progressive Farmer's Institute, which provided up-to-date information and educational programs for rural residents. In 1897, the institute supported compulsory education, and in the early twentieth
century it pushed for rural free delivery and a modern system of
roads. As a young man in his twenties, Snow gave lectures on farm

14. James Whitcomb Riley, *The Complete Poetical Works of James Whitcomb Riley,* 436.

15. Revell, *James Whitcomb Riley,* 31; *Poplar Bluff (Mo.) Daily American Republic,* January 17, 1955.

16. *St. Louis Post-Dispatch,* January 16, 1955; Snow, *From Missouri,* 62–77.

17. Snow, *From Missouri,* 65–69.

economics and tariffs. These lectures and his work with the Farmer's Institute marked his first public appearances as an analyst and critic of the economic system as it applied to agriculture. Late in his life, he expressed the belief that these activities marked his emergence as a "class-conscious farmer." With less exaggeration, he reported that his association with this progressive group and his opposition to tariffs led him away from his father's party and into the fold of the Democrats, who represented the majority of citizens in Hancock County.[18]

When Snow left for Missouri, Greenfield was the thriving hub of a prosperous county. The town boasted several department stores, at least eight grocers, two hotels, two daily and several weekly papers, and a lively variety of churches, fraternal orders, literary societies, and social clubs. His father was a prominent citizen who had won election as county chairman in 1878, 1880, 1882, and 1884, and county recorder in 1886.[19] Thad Snow had a substantial farm, a family, and enough status on his own to be elected president of the Farmer's Institute.

Telling his story at the end of his life, he said that he left all this behind in response to an unaccountable feeling of restlessness and a perception that Greenfield and central Indiana were "too complete and well ordered, and had not enough left that required doing."[20] He looked first at a plantation in Mississippi, but the land deal fell through. On the train back to Indiana, he met a man who told him about the rich land being drained on the west side of the Mississippi River. He and his new acquaintance got off the train at Cairo, Illinois, where the Ohio and Mississippi Rivers converged, and took a ferry to the Missouri side.

There, Snow presented himself to a land salesman in Charleston, just twelve miles from Cairo, a town that suffered frequent floods. The lowlands on the west bank of the river flooded numerous times in the late 1800s, but the land salesman assured Snow that this could not happen again because of the new levees in Mississippi and Scott Counties. Snow had enough native sense to doubt this. But the threat of flood did not deter him. With visions of a vast,

18. Ibid., 19, 78, 82–86.
19. Bowen, *Biographical Memoirs of Hancock County,* 595–699.
20. Snow, *From Missouri,* 90.

wooded country at the edge of a dangerous river, he went back to "civilized Indiana" to make a decision. "I did not come to my senses," he wrote. "Instead I returned to Swampeast Missouri and bought that insurmountable tract of wild land." He bought the land in late fall of 1910 and moved his family to the farm in spring of 1911.[21]

Looking back on his move to Missouri, Snow questioned both his motives and his judgment. He might have bought a smaller tract of cleared land, but he wanted large acreage, and even more than that, he wanted the chance to conquer a big, wild, wooded piece of ground. As he expressed it:

> I had the big-eye. I was a fool, looking for trouble, punishment, and perhaps disaster. The improved farms were much the cheapest, really. All the uncleared land was priced much too high and could be sold because outside buyers and even local buyers, who ought to have known better, always figured they could clear and improve land for about half what it would inevitably cost in the end.[22]

He had the gambler's weakness for the long shot that just might pay off, and he had no desire to play it safe.

With pioneer gusto, he cleared the trees from his land. As much as he liked to view himself as a self-reliant frontiersman, this process involved hiring displaced families to do the backbreaking work of cutting and hauling the timber. Like other landowners, he paid them by the acre. In his memoirs, he expressed admiration for these anonymous workers, calling them the "valiant axemen and axewomen of the lowlands" and depicting them as hardy pioneers. But he also confessed that they received little reward for their "monumental labor," with wages amounting to less than forty cents a day.[23]

Once cleared, the land needed draining. To do this, Snow hired workers to dig ditches, or culverts, and lay down clay tiles.[24] The tiles routed surface water away from the fields to drainage ditches

---

21. Ibid., 90–97.
22. Ibid., 94.
23. Ibid., 134.
24. Snow's granddaughter described this method of draining the land. Some of the tiles were gone in 2002, but many remained (Corse interview, May 28, 2002).

and finally to streams that fed into the Mississippi River. The clearing of thousands of acres in this manner, of course, had dramatic consequences for the lowlands, creating a need for massive flood-control projects to protect the cleared land.

For Snow, there were financial consequences. His daughter Fannie recalled that "Dad always said he paid as much for the tile under ground as he did for the farm itself."[25] He funded drainage and developing the land by borrowing from insurance companies. For instance, in 1916, he gave the Prudential Insurance Company of America a claim to all rents on his land. One year later, after he paid off his debt, the insurance company released its claim. However, in that same year, he obtained another mortgage from Phoenix Mutual Life Insurance Company. Despite these complications, he remained ambitious. In 1919, he acquired an additional tract of about 450 acres adjacent to his original farm.[26]

In many ways, at least until the late 1920s, he was the quintessential developer and modernizer, boosting the local economy and dragging the region into the twentieth century. In the early 1920s, he served as president of the board of directors of the Southeast Missouri Agricultural Bureau, an organization dedicated to advertising and attracting new populations to the region. His friend Xenophon Caverno also served on the board.[27]

Under Snow's leadership, the bureau maintained offices and exhibit rooms in St. Louis's Union Station, displaying the agricultural products of the Bootheel. Members of the bureau made connections with the Memphis Chamber of Commerce, inviting "such cooperation from that city as will enable further activity along development lines."[28] Under the board's direction, various committees lobbied state and federal governments to promote reclamation work in the lowlands.

25. Fannie Snow Delaney, interview with the author, Snow's Corner, Mississippi County, Mo., February 9, 1996.

26. Mississippi County, Mo., deed book 70, p. 559; Mississippi County, Mo., deed book 77, p. 181; Mississippi County, Mo., deed book 59, p. 476; deed book 98, p. 505; Mississippi County, Mo., deed book 94, p. 8.

27. *Sikeston (Mo.) Standard,* May 9, 1922. Other members of the board were A. I. Foard of St. Louis, Judge C. A. Vandivort of Cape Girardeau, W. H. Sikes of Sikeston, S. P. Reynolds of Caruthersville, Dwight H. Brown of Poplar Bluff, Norman D. Blue of Puxico, and R. Irl Jones of Kennett.

28. Ibid.

Snow campaigned vigorously to bring highways to the Boot-heel, making numerous trips to the state capital to lobby for road and bridge projects. He later recalled that he "drank likker and did boosting in the good old days" with highway commissioners and engineers and helped to bring the first concrete roads to Mis-sissippi County. He had direct access to a high-ranking official when his Scott County friend Charles D. Matthews became a Re-publican member of the State Highway Commission in 1921.[29] Reappointed in 1924, Matthews became chairman of the commis-sion in 1926.[30]

When Snow first arrived in Missouri, the state was just beginning to play a role in road building. In 1907, the General Assembly passed a law providing for the appointment of a State Highway Engineer by the Board of Agriculture. Two years later, the legisla-ture established a general road fund. Federal funds for highway building became available in 1916, and one year later Missouri gained access to these funds by passing the Hawes Road Law and establishing the bipartisan State Highway Commission. The Centennial Road Law of 1921 formally created a state highway sys-tem and initiated widespread construction projects using funds from a sixty million dollar bond issue approved by voters in 1920. Within the first two years, Mississippi County benefited from major road projects utilizing local, state, and federal monies totaling more than eight hundred thousand dollars. The most important project connected Bird's Point to Charleston and continued on to Sikeston, providing a section of Highway 60, linking Cairo, Illinois, and Poplar Bluff, Missouri.[31]

In 1928, with motor vehicle owners insisting upon a more rapid completion of the state's road system, the Automobile Club of Missouri campaigned for a second authorization of road bonds. With Matthews as chair of the Highway Commission, which sup-ported Proposition 3, a constitutional amendment allowing the state to issue fifteen million dollars of road bonds every year for five

29. Snow Papers, folder 50; State Highway Commission of Missouri, *Sixth Biennial Report* (Jefferson City, Mo., 1928), [2].
30. Ibid.
31. State Highway Commission, *Second Biennial* Report, 76, and *Third Bien-nial Report,* 8–10, 103–6.

years, beginning in 1930. The proposition passed, becoming effective in March 1929.[32]

Snow opposed this second bond issue in a public letter to T. H. Cutler, the Highway Commission's chief engineer. Expressing "a regret that amounts to a real sorrow," he stated that he hoped Missouri voters would defeat the proposed constitutional amendment dedicating gas and license tax fees to road building and authorizing a bond issue for new highway construction. He quarreled with the concept of "government by amendment," believing that the legislature and the commission should have the flexibility to make decisions according to changing needs and circumstances. He believed that the 1928 proposal favored road construction in urban rather than rural areas. In the interests of providing roads to "backward areas," he was willing to accept a delay in the reconstruction of Highway 61 between Cape Girardeau and Portageville, believing that the existing highway could carry traffic for another ten years. In general, he questioned the fairness of the measure, although he hoped that by opposing the initiative he would not "forfeit your [Cutler's, and certainly also Matthews's] regard."[33]

Snow's strained relationship with highway officials continued through the next decade. In November 1939, he visited the governor of Missouri and expressed the opinion "that the Highway Department [was] going to pot."[34] In particular, he believed members of the Highway Commission, including H. G. "Chilli" Simpson of Charleston, had an apparent conflict of interest because of ties with the petroleum and asphalt industries. This was just one example of the ways in which Snow became disconnected from powerful businessmen in the Bootheel.

His waning enthusiasm for road building signaled a deeper change in his thoughts and attitudes. Looking back on this time, and generalizing his own personal feelings to include all the people

32. State Highway Commission, *Sixth Biennial Report,* 50, and *Seventh Biennial Report,* 34–35.

33. "Thad Snow Opposes Road Bond Issue," April 5, 1928, clipping in scrapbook, Snow Papers.

34. Thad Snow to Hon. Lloyd C. Stark, November 15, 1939, folder 2578, Governor Lloyd Crow Stark Papers, Western Historical Manuscript Collection, University of Missouri–Columbia.

of Southeast Missouri, he wrote, "In 1920, we were confident, buoy-ant, and aggressive; in 1930 we were defeated, confused and list-less."[35] Snow began the decade as an unembarrassed economic booster, happily fraternizing with big businessmen and Republi-cans and vigorously championing the development of modern highways. As he entered the next decade, he regretfully loosened ties with old friends and questioned the headlong rush to build roads to a clouded future.

His personal circumstances changed dramatically in the 1920s. As he entered his fifties, with a new wife and two young daughters to support, he faced a financial crisis. In 1923, after he defaulted on a loan and failed to pay taxes, the International Insurance Company took title to the 450-acre tract he had purchased in 1919.[36] Four years later, Snow's brother-in-law, Lawrence A. Simpson, had pos-session of the tract and sold it back to Snow for one dollar, with Snow assuming various liens.[37] Snow defaulted on his obligations, and in the dismal month of October 1929, he conveyed the land in trust to a creditor to secure his mounting debts.[38] The abstract of the property indicates that Snow defaulted again and lost the title to this tract. However, in the summer of 1933, his brother-in-law came to the rescue, acquired the property, and sold it back to Snow and his wife, Lila, for "one dollar and love and affection."[39]

These were difficult times for the family. His daughter Fannie re-membered that during the early 1930s she had to go to the country school, because the tuition in Charleston was too high. At one point, she remembered going barefoot to school, but she was not the only one. The Great Depression had reduced many families to these cir-cumstances. "It was the norm," she said. In order to cover his debts, her father had to cut down and sell more of the cypress trees on his

35. Snow, *From Missouri*, 153.
36. Mississippi County, Mo., deed book 94, p. 8.
37. Mississippi County, Mo., deed book 100, p. 402.
38. Mississippi County, Mo., deed book 103, pp. 324–25, and deed book 104, p. 426.
39. Mississippi County, Mo., deed book 105, p. 502. Ms. Hazel Williams of the Mississippi County Abstracting Company provided access to the abstract of the property, which indicates that Snow subsequently conveyed the property to his daughter Emily. After her death in 1948, her heirs (Hal Snow and Frances Snow Delaney) conveyed it back to Thad Snow.

land. "It really hurt him to do that. But he needed to pay bills. He was just trying to make it."[40]

The family made it through these hard times. A reporter who visited him in early May 1937 concluded that Snow's "thousand acres of rich bottoms land [sic] would flourish if the rest of Missouri suddenly became desert." Hired hands cultivated the corn and tended the livestock. Sharecroppers worked about three hundred acres of cotton, and even in the worst years of the Depression, his land consistently produced more than the "bale-an-acre yield most planters dream about."[41]

In 1937, Snow was fifty-five, tall and lean, with short-cropped gray hair, a suntanned face, red-tipped nose, and white-rimmed glasses. His daughter Fannie was fifteen; her younger sister, Emily, was eleven. The reporter found them charming. Their mother, he said, was a "sympathetic wife," who put up with her husband following her around the house and reading aloud from Thorstein Veblen, whose most famous work was *The Theory of the Leisure Class.*[42]

Snow's enthusiasm for Veblen appeared to be that of a new convert. Exactly when Veblen's ideas entered his consciousness remains uncertain. Before 1910, he had become active in a progressive farmers' organization and turned away from the Republican Party. In the 1920s he had behaved like a typical booster, pumping for highways in the Bootheel. By the end of that decade, faced with financial problems, he questioned the headlong rush toward development and modernization. The Depression jolted him. His farm remained productive, but the reporter who visited him in 1937 found a man looking for an explanation of what had happened to the American economy.

The reporter was observing a man unable to help his ailing wife, a man coming to grips with private pain as well as nationwide turmoil. Since their courtship, Thad and Lila had graduated from the

40. Delaney interview, February 9, 1996.
41. *St. Louis Post-Dispatch,* May 2, 1937.
42. The reporter visited just a few weeks before Lila's death. She had been in poor health for more than a year and had undergone several operations, but she deteriorated rapidly in May 1937 and died suddenly on May 17 (*Charleston [Mo.] Enterprise-Courier,* May 20, 1937).

intensely personal love poems of Elizabeth Barrett Browning to the caustic social criticism of Veblen. For Snow, although he remained a family man, this may have represented a turning away from intimacy toward a concern with the wider world. It may also have indicated an examination of his own conscience.

When he first came to the Bootheel, Snow exhibited some of the characteristics of the "barbarian temperament" that Veblen derided in the American elite. A white male with money in his pocket, Snow was a risk taker with a strong belief in his own competence and a naive faith in the power of luck. While he was neither unscrupulous nor bellicose, he demonstrated the acquisitive temperament and the gambling propensity that were hallmarks of the powerful males whom Veblen skewered in his classic work of social criticism.[43]

In many ways, however, Snow differed from Veblen's archetype. For instance, he had no formal or informal ties to any religious groups, or in Veblen's terms, "anthropomorphic cults." A youthful conversion experience failed to bind him to the church. As a farmer in central Indiana, he had had many Quaker neighbors. He respected their personal integrity and their strong faith, but he remained an agnostic. Long after his death, his daughter recalled that he never attended church, except when he visited his friend Owen Whitfield, who preached in African American churches in the Bootheel, Cropperville (Butler County), and Cape Girardeau, Missouri, and Du Quoin, Cairo, and Mounds, Illinois.[44]

Unlike Veblen's wealthy conformists, Snow had little fear of social disapproval. When he suffered from corns as a young farmer, he defied convention by going barefoot, even in the snow.[45] When he first came to Missouri, development and drainage had begun, but the land was so swampy and undesirable that people assumed he and other newcomers had left their old homes "under a cloud and pursued by the Sheriff."[46] He seemed to enjoy this tinge of dis-

43. Thorstein Veblen, *The Theory of the Leisure Class*, 225, 246.

44. Ibid., 293; Delaney interview, August 16, 1999; Lawrence O. Christensen, William E. Foley, Gary R. Kremer, and Kenneth H. Winn, eds., *Dictionary of Missouri Biography*, 792–93. Dan Cotner recalled that, when Whitfield was a minister in Cairo, Illinois, he had to come to Cape Girardeau to find a dentist who would accept black patients (interview, April 4, 2002). Shirley Whitfield Farmer said her father was a pastor in Du Quoin and Cape Girardeau in the 1950s and at Mounds, Illinois, in the early 1960s (interview, April 7, 2002).

45. Snow, *From Missouri*,

46. *St. Louis Post-Dispatch*, May 2, 1937.

reputability. Later in life, he accepted and even relished the epithet "devil of the Bootheel" as his well-earned reward for standing up for the downtrodden and defiant sharecroppers.

He never identified himself as a member of the social group that Veblen castigated as the leisure class. During his early years in the Bootheel, he enjoyed spending time at the mule sales barns, socializing with the dealers and farmers, large and small, rich and poor. Like the western cowboys, he took pleasure in breaking horses and mules to the bridle. Mules, especially, appealed to him, because of their tenacity and amazing strength. With tongue in cheek, but also with fondness, he wrote, late in his life, "I have actually, a few times, caught a mule's hind hoof in my hand and shaken it, as I would yours, because it was just the mule's way of saying, 'How do you do?' "[47]

He liked to remind people that he was a real farmer who got his hands dirty and put his back into the work of tilling the soil. But in reality, he profited from the work of hired hands and sharecroppers. By inviting the union to organize his workers and by supporting the sharecroppers' strike in the 1930s, he would demonstrate that this reality troubled him.

Admittedly ambitious, he never succumbed, however, to a desire for luxury, or in Veblen's words, conspicuous consumption. Decades after settling in the Bootheel, and after his farm reached a peak of productivity, he continued to live in a modest house. An observer described it as "a tiny shingled cottage not much bigger than his croppers' cabins."[48] A friend of his daughters' remembered that as late as the 1940s there was no electricity or running water in the house. Snow took a shower in a tin stall with a bucket attached to a pulley.[49]

While he did not build an ostentatious house, he did purchase and develop a big farm, and he wanted to make it profitable. In order to achieve his goal, he exploited nature. He tried to save trees on a portion of his farm, and it bothered him to sell cypress logs during the Depression. But at the end of his life he was still capable of writing nostalgically of the way in which he and Mose Feezor dragged huge cypress logs off the cutover fields, in spite of the damage

47. Snow, "Proud Kate," 65.
48. *St. Louis Post-Dispatch,* May 2, 1937.
49. Warren interview, November 7, 2002.

they had inflicted on the environment. Viewed in the most simplistic terms, this was a victory of man over nature.[50] Nevertheless, in his declining years he retreated to the forests of the Ozarks and eloquently described the beauty of the wooded landscape.

Like other Bootheel entrepreneurs, he exercised dominance, not only over the land but also over the people who worked for him. Although he shared a genuine friendship with Feezor and showed kindness toward his family, Snow remained aloof from them. Other hands came and went. When African American sharecroppers moved to his farm in the 1920s, he kept his distance. The newspaper reporter who interviewed him in 1937 asked him if it was possible to make money farming under the existing economic system. With irony and self-criticism, he replied, "Sure it is. All you have to do is work plenty of rich bottom land—and pay your workers as little as possible."[51]

He knew there was something terribly wrong in his world. Life in the Bootheel chastened him. If he came there to be close to nature, he knew that he had altered the natural world irreversibly—and not for the better. If he came there to make a fortune, he had experienced some hard knocks. But his own worries paled in comparison to the suffering of people around him.

If he viewed himself as a hardy pioneer, he had to wonder if perhaps the real pioneers were the timber cutters and sharecroppers who came to this region, just as he did, looking for a new start. In 1910, he decided that he "was going somewhere to a different sort of country."[52] But it turned out to be different in ways he never imagined.

50. Snow, "Proud Kate," 68.
51. *St. Louis Post-Dispatch*, May 2, 1937.
52. Snow, *From Missouri*, 90.

Flood Culture 3

If people of the Delta lowlands are different from people elsewhere; if their "culture" is distinctive and therefore interesting, that is at least partly because the waters of the Mississippi and of the lesser streams coming down out of the Ozark hills have spread over so much of the lowlands so often in the past, and in some large areas still do so or threaten to do so during the winter and spring of each year.

—Thad Snow, *From Missouri*, 98

Description can give no sense of the dread realities of flood misery—the cold mud, the lost goods, the homeless animals, the dreary standing around of destitute people.

—Thomas Hart Benton, *An Artist in America*, 146

In 1937, the citizens of Mississippi County postponed their centennial celebration because of devastation due to flooding.[1] Thirty-five years later, local historian Betty F. Powell recounted a long saga of flooding in the county. Pioneer Abraham Bird left a monument showing the high-water mark at fifty feet in 1815. County residents took a perverse pride in recording such disasters. Powell quoted the headline in the Charleston newspaper's centennial edition, which boasted that "EARTHQUAKES, FIRES, CYCLONES AND FLOODS FAIL TO HALT PROGRESS OF PIONEERS."[2]

In 1882, the river topped its 1815 mark by fourteen inches, inun-

1. *Cape Girardeau Southeast Missourian*, May 11, 1937.
2. Powell, *History of Mississippi County*, 32.

dating all the farms at Bird's Point, Wolf Island, and many other settlements. While they were proud enough to record the event, residents clamored for more and better levees. Despite some levee building, a flood in 1897 nearly destroyed Bird's Point, washing away fifteen houses and driving frightened residents to seek refuge in boxcars. Looters trawled in boats, stealing furniture and other property. A Wolf Island resident struggled to find humor in the situation, telling the local paper that instead of tending the crops, farmers were gigging fish in their fields.[3]

By the time Snow purchased his land, Mississippi County residents had raised hundreds of thousands of dollars by private subscription and taxes to create a system of levees and ditches that would divert water away from the lowlands. In 1910, landowners began using drain tiles to carry off excess water. An enterprising resident named J. I. Belote operated a concrete tile plant in Charleston. When he came to the county, Snow would join other farmers in the expensive process of digging ditches and laying tiles to drain the land.

Despite these efforts, Snow and his neighbors continued to suffer from periodic inundations. Floods in 1912 and 1913 demonstrated the power of a great river, but the great floods of 1927 and 1937 changed the course of history in the lowlands. After 1927, the federal government invested heavily in a system of levees and floodways that protected the commercial centers and farmlands of southeastern Missouri and Illinois. The system included a spillway between the main levee and a setback levee in New Madrid and Mississippi Counties.

In 1937, the government made the decision to breach the outer levee and flood out the sharecroppers in the spillway east of Charleston. Snow stood on high ground behind the setback levee that protected his land and watched poor farm laborers escape the swirling waters with a few belongings, their families, and their lives.

Snow remembered the flood of 1912 with a feeling of nostalgia for his early pioneer days. When high water threatened, he and his hired hands piled sandbags on top of a county-built levee. The levee gave way two miles north of his house. "I woke up in the morning," he wrote, "to hear the roar of the incoming water and see it eddying muddily in a draw fifty feet from my door." Long after

3. Ibid., 32–36.

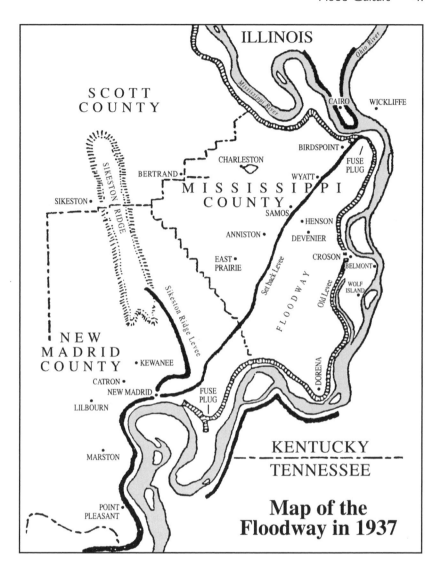

**Map of the
Floodway in 1937**

the event, he described the process of building boats and hauling
livestock to dry ridges and native American burial mounds. What
struck him most forcefully was the cheerful, hearty way in which
his neighbors coped with the calamity.[4]

Another Mississippi County pioneer remembered putting his

4. Snow, *From Missouri,* 98, 101.

furniture up on scaffolding and sleeping on a raised platform while the water seeped up through the floorboards of his house. When houses flooded, farmers fled to their barns, William A. Wyatt recalled. The government sent large boats to rescue the victims.[5]

According to Snow, the homeless found lodging with relatives, friends, or kind strangers. True to the cult of rugged individualism, "People were supposed to take care of themselves." Without a sense of mutual responsibility, the flood victims maintained what Snow described as "a fine sense of unity."[6] At this time in Mississippi County, he recalled that the victims were relatively few, mostly white, and not yet separated by obvious differences in wealth and status.

Snow's memory may have been too rosy. On April 5, 1912, the city of Charleston was high and dry, but thousands of homeless people sought refuge there from the east, north and south, after the Big Lake levee broke. People who were unable to find shelter received temporary relief from the city and the county. Many had to leave livestock, goods, and clothing behind when they fled their homes. Others were stranded in their houses or on high mounds surrounded by floodwaters. Residents of Charleston and East Prairie went out in boats to rescue people, some of whom had climbed out onto their roofs. According to the newspaper, "Some of the people driven to Charleston and who are being cared for by our people were very poorly clad, the little children not having enough clothing to keep them warm, no shoes and crying for something to eat."[7]

The situation seemed much more desperate than the one Snow remembered. Nevertheless, he may have correctly judged the buoyant spirit of the people. The newspaper reported: "It is a pitiful sight, yet the unfortunate people take things very easy and are not worrying half as much about their condition as those who are caring for them."[8]

Damage from the 1912 flood was very costly. According to the Charleston paper, farmhouses, barns, and outbuildings were wrecked and washed away. Livestock drowned, wheat fields were destroyed, and lumber was scattered far and wide. Losses of property and live-

5. Powell, *History of Mississippi County*, 41.
6. Snow, *From Missouri*, 102–3.
7. *Charleston (Mo.) Enterprise-Courier*, April 5, 1912.
8. Ibid.

stock amounted to at least a million dollars. Repairs on the Big Lake levee would cost the county at least twenty thousand dollars. The Mississippi River Commission, the Red Cross Society, and private donors assisted with flood relief.[9]

Wyatt recalled two conflicting reactions to the 1912 flood. His father and members of the older generation insisted that there could never be such a terrible flood again. But many farmers learned to expect the worst, quickly rebuilding levees and obtaining boats in case of another disaster. Living with the threat of floods fostered both a brave optimism and the practical impulse to prepare for adversity.

When floods came again in 1913, county residents heeded warnings to evacuate their families and their livestock. Wyatt remembered that when the government gave out a flood warning in the spring of that year farmers moved their animals to higher ground. Despite these precautions, the 1913 flood took its toll on the free-ranging stock.[10]

After the waters receded, Snow bought a new herd of young cattle and tagged them. By mistake he tagged sixty head that did not belong to him. He later corrected the error, but he recalled that one of his neighbors continued to view him as a cattle thief. A few years later the neighbor, H. C. Pierce, sued him for falsely claiming ownership of a black heifer and her red and white spotted calf. In July 1917, the court found for the plaintiff in the case. Snow appealed, but his appeal was denied.[11]

In his memoirs, he wrote that the flood of 1913, the second in two years, was one too many, "And nobody liked it."[12] The camaraderie of pulling together during a natural disaster could easily wear thin. After 1913, the federal government undertook massive levee-building projects, and this was the last time Snow's acreage flooded.[13]

Snow's Corner escaped inundation in the spring of 1927 when both the Ohio and the Mississippi Rivers overflowed, causing one of the worst disasters in U.S. history. During this flood, most of

9. *Charleston (Mo.) Enterprise-Courier,* April 5 and 12, 1912.

10. Powell, *History of Mississippi County,* 41.

11. Ibid., 108–12; Circuit Court Record, Mississippi County, Mo., book 26, pp. 353, 367.

12. Snow, *From Missouri,* 105–6.

13. Delaney interview, March 10, 2001.

Mississippi County's levee held, but the dikes broke at Dorena, just north of New Madrid, in the southern tip of the county. Below this point, river waters covered much of the delta from Cairo to the Gulf of Mexico. The flood caused distress in seven states, but most severely in Mississippi, Arkansas, and Louisiana.[14]

During the worst months of the flood, the U.S. Army Corps of Engineers (Corps), the American Red Cross, and other agencies responded with a truly heroic campaign to provide shelter and distribute food to victims. The Corps provided boats, equipment, and quarters for Red Cross volunteers, Coast Guard rescue teams, flood relief workers and refugees. The U.S. snag boat *Macomb*, originating in St. Louis, traveled down the Mississippi River and became the depot boat for the relief fleet and served as a base for the distribution of Red Cross supplies. The deck crew worked day and night, transferring supplies and repairing watercraft. Engineers patrolled levees in motorboats. The *Macomb* left St. Louis on April 15, arrived in Cape Girardeau on April 25, headed for Memphis on April 27, and returned to St. Louis on June 25, after the main crisis had passed, although the waters continued to recede very slowly.[15]

The most important result of this flood for Mississippi County was Major General Edgar Jadwin's plan for flood control. Jadwin, the chief of the Corps, called for a series of floodways paralleling the general course of the river.[16] In southeastern Missouri, the plan called for the creation of the New Madrid Floodway, encompassing 131,000 acres of farmland from Bird's Point in northern Mississippi County to the city of New Madrid, thirty-five miles to the south.[17]

The purpose of the New Madrid Floodway was to create a safety valve that would protect Cairo and the cities and farms downriver.

14. Joe Moore's scrapbook on the 1927 flood, on file at the offices of the *Charleston (Mo.) Enterprise-Courier,* and Pete Daniel, *The 1927 Mississippi River Flood,* 9.

15. National Archives and Records Administration, Kansas City Branch, RG 77, Records of the Office of the Chief of Engineers, St. Louis, Missouri, District, General Correspondence, box 26, folder 7410.

16. Edgar Jadwin, Major General, Chief of Engineers, "Flood Control of the Mississippi River and Its Alluvial Valley" (Washington, D.C.: War Department Office of Chief of Engineers, December 1, 1927), 4, filed in Record Group 77, box 24, folder 7402.

17. Edison Shrum, "Super Floods Raging in Wide Spread Area: The *Scott County Democrat*'s Account of the 1937 Mississippi River–New Madrid County Jadwin Floodway Disaster," 1.

According to Jadwin, the most serious flooding problem began at Cairo, on the east side of the river, just north of Bird's Point, at the confluence of the Ohio and the Mississippi. In Jadwin's words, "From here to New Madrid the main levee on the west bank chokes the river unduly and should be set back sufficiently to lower the head of the water at Cairo by six [feet] in an extreme flood."[18] In order to relieve this problem, he proposed that the existing levee be lowered five feet. The Corps would then build another levee, called a setback levee, farther away from the river. The floodway between the old and new levees would be usable as farmland, except during big floods.

In the event of a "superflood," the engineers reasoned, water would flow over the old levee, breach it, and fill the five-mile-wide basin between the river and the setback levee.[19] By allowing water to enter the floodway, engineers hoped to reduce the pressure and destructiveness of floodwaters headed for Cairo and other cities downriver. At a point east of New Madrid, where a peninsula jutted out into the river, the Corps proposed to create a "fuse plug": If the river topped the levee, water would flow into the floodway and then exit through the fuse plug into the main channel.[20]

Snow's friend Lucius T. Berthe, a resident of Mississippi County and consulting engineer for the county's Levee District Number 3, passionately opposed the plan. He realized, as Jadwin's report plainly stated, that the plan's main purpose was to protect the fifteen thousand residents of Cairo. Berthe protested loudly that the plan would "crucify Missouri on the cross of Cairo's necessity, yet fail to provide for her salvation."[21] Berthe argued to Congress, the courts, and anyone who would listen that the Corps' calculations were wrong and that the river would not behave as predicted in the event of a great flood. For the next ten years, he continued to defy the Corps, gathering facts and figures challenging their decision.[22]

Booteel farmers voiced deep concerns about the Jadwin plan but failed to block its completion. Scott County historian Edison Shrum recalled that a considerable number of families had homes and

18. Jadwin, "Flood Control," 6.
19. Shrum, "Super Floods," 2–3.
20. Jadwin, "Flood Control," 54.
21. *St. Louis Globe-Democrat,* June 9, 1929.
22. Snow, *From Missouri,* 216.

farms in the New Madrid Floodway. The plan to provide a safety valve for the city of Cairo held little to appeal to them. "In fact they were mad as hell about it and fought it in every way they could."[23] They held protest meetings, appealed to Congress, and tried to negotiate with the Corps of Engineers. But negotiations failed.

The Corps built the setback levee between 1929 and 1931. The levee, which required more than eight million cubic yards of earthwork, began at Bird's Point and was set back five miles west of the main levee along the river. Mississippi County received forty thousand dollars for lands required for the right of way of the setback levee and a drainage canal. Another thirty-six thousand dollars went to state road bonds.[24] Under the plan, the government also paid individual landowners for the right to breach the main levee and flood their land, but the parties could not readily agree on property values. In March 1930, five hundred citizens of Mississippi and contiguous counties attended an acrimonious meeting with federal officials. These officials accused county land appraisers, "actuated by prejudice against the government," of inflating values. Landowners protested that the flood project was designed to protect Cairo, not southeastern Missouri.[25]

Snow did not take an active role in the protests, judging the cause to be hopeless, but he believed the Jadwin plan cruelly sacrificed Mississippi County in order to save the cities and lands downriver. He remembered it this way:

> The whole thing was wrong. It was maliciously wrong. Or so we felt at the time; our judgment of the equities involved may have been clouded by our passion. Anyway we hated General Jadwin and the town of Cairo with a great and holy passion. But all our hot words and factual arguments came to precisely nothing, unless they served to harden the hearts of the Engineers against us and caused them to treat us rougher in many respects than they otherwise would have behaved toward us. They did treat us rough.[26]

23. Shrum, "Super Floods," 2.
24. Powell, *History of Mississippi County*, 43.
25. *Cape Girardeau Southeast Missourian*, March 11, 1930.
26. Snow, *From Missouri*, 204.

Most people believed that, in the event of a major flood, the Corps would simply allow floodwaters to top the main levee and flow into the setback basin. As a matter of fact, the Corps planned, if necessary, to breach the main levee with dynamite.

Edison Shrum recalled that proponents of the plan insisted that breaching the main levee and filling the flood basin would lower the water level at Cairo by at least six feet. They also asserted that the water level within the floodway would not top eight feet and, therefore, would not destroy houses. Both these predictions proved disastrously wrong.[27]

In January 1937 massive overflows of the Ohio and Mississippi Rivers caused damage and suffering in one fourth of the states in the nation. Rising very suddenly in bitter wintry conditions, the Great Depression–era flood affected nearly two hundred counties in twelve states from Pennsylvania to Louisiana. Nearly a million people had to leave their homes because of the flood. In Louisville, Kentucky, more than half the city's population vacated flooded neighborhoods. In Cincinnati, Ohio, oil tanks collapsed, causing dozens of fires. In Paducah, Kentucky, officials ordered thirty-five thousand people to evacuate.[28] On January 16, levees broke along the St. Francis River in southeastern Missouri. By the end of the month, record flood levels on the Mississippi River threatened to inundate the fourteen thousand residents of Cairo.[29]

Clearly, this was at least in part a man-made disaster. For decades, loggers had cut over the forests along the Ohio and Mississippi Rivers and their tributaries. Farmers had cleared the land, killing the brush and smoothing away the rough places that slowed water as it flowed toward the streams. Drainage projects had moved the lazy water of swamps into ditches and channels that emptied into swiftly flowing streams. When unusually heavy rains fell, tributaries filled rapidly and poured their loads into the big rivers.[30] The levee system constructed by the Corps of Engineers protected the river basins against seasonal high water but not against a monumental rise. The higher the levees, the greater the catastrophe.

27. Shrum, "Super Floods," 3.
28. *Scott County (Mo.) Democrat*, January 28, 1937.
29. Joe Moore's scrapbook.
30. Daniel, *1927 Mississippi River Flood*, 7.

The result in 1937 was the "superflood" envisioned in the Jadwin plan. The time had come to relieve the pressure on Cairo by inundating the lands between the main levee and the setback levee in the New Madrid Floodway. Breaching the main levee at this point would reduce the height of the floodwaters downriver, but it would have a devastating effect on the farmlands in the floodway.

On Monday, January 25, the army used two thousand pounds of dynamite to blow the main levee. Blasting the levee was at once the arrogant outcome of human calculation (or miscalculation) and a humble admission that human intelligence had failed to control the ultimately ungovernable power of nature. Snow heard the explosions from his home twelve miles away. There was little or no advance warning. As Snow recalled, "It didn't occur to us that the Army would blast the levee open with dynamite, and certainly they would not do a thing like that without notice in advance to everybody concerned. But they did exactly that."[31]

Shrum, living fifty miles from the levee, heard and felt a series of "tremendous blasts" and felt vibrations or tremors on the afternoon of January 26. He speculated that the earthquakelike tremors resulted from the monstrous weight of the water that suddenly rushed into the floodway. When he wrote to the Corps of Engineers office in Memphis, Tennessee, inquiring about the matter, he received a brief reply, stating that "this office does not feel at liberty to furnish the information requested, as it is forbidden by law to furnish information which may in some manner be used as the basis of a claim against the Government."[32]

The army estimated that 2,500–3,000 refugees left the spillway in the two days before the breach. In a telegram dated January 27, 1937, the day after the breach, Major Roy D. Burdick of the Corps of Engineers asked Lieutenant Colonel P. S. Reinecke, the District Engineer in St. Louis, to send clothing and equipment for 2,500 refugees in Charleston.[33] Because of inadequate warning, many residents of the floodway had failed to evacuate before the breach. The Highway Department had to send rescuers in motorboats.[34] Many peo-

---

31. *Cape Girardeau Southeast Missourian*, January 25, 1937; Snow, *From Missouri*, 210.
32. Shrum, "Super Floods," 3–4.
33. National Archives Record Group 77, box 26, folder 7413.
34. *Cape Girardeau Southeast Missourian*, January 25, 1937.

ple walked or tried to escape in wagons while floodwater sloshed up around their feet and thick, roiling mud engulfed the wagon wheels. Others went back into the basin, risking their lives, to try to salvage livestock and personal possessions.

The economic cost was severe. Damages in the New Madrid Floodway far exceeded a million dollars. A federal government report issued in March 1937 indicated that fifteen hundred farmers suffered property damage when nearly 140,000 acres of land was inundated. Floodwaters destroyed five hundred buildings and killed 50 percent of the livestock in the basin.[35] At the time of the flood there were six villages in the basin, including Bayouville, Belmont, Crosno, Dorena, Medley, and Wolf Island. Only two of these communities, Dorena and Wolf Island, survived.[36] The people in the other communities lost not only their livestock and crops but also their homes.

Snow compared the exodus from the flooded area to the displacements of war. As in a war, no one was certain what would happen next. The refugees had to evacuate in a blinding sleet storm with floodwaters boiling over the levee behind them. Snow had seen previous evacuations in rain and mud, but he had never before seen one in six inches of frozen sleet. Many of the evacuees drove mules through five miles of standing water in an icy wind. Many had their hands and feet frozen. On the highway that passed by his home, Snow witnessed "an unbroken line of traffic," consisting of exhausted cattle, mules, and people carrying a small amount of bedding or their children on their backs. Most of the refugees were tenants, sharecroppers, or day laborers.[37]

Following the debacle in the floodway, the army hastily recruited men to shore up the setback levee. On Saturday, February 11, a barge crammed with more than a hundred workers sank in the roiling ice water. Thirty men drowned in the disaster, which the coroner's jury blamed on negligence due to "hurried and inadequate" preparation for dealing with "an emergency in the flood situation."[38]

Throughout the last week of January and the first weeks of February, refugees poured out of the flooded river bottoms. On January

35. *Scott County (Mo.) Democrat*, March 25, 1937.
36. Shrum, "Super Floods," 9.
37. Snow, *From Missouri*, 214–22.
38. *Scott County (Mo.) Democrat*, February 11, 1937.

28, the *Scott County Democrat* reported that five hundred families sought refuge in New Madrid. As rain and seep water filled the streets there, rescuers had to move the refugees. Displaced farmers from the Charleston area sought refuge in the Scott County Courthouse in Benton. Trains stopped running and roads were closed. Rescuers depended on boats, which were in short supply.[39]

The Red Cross, the Salvation Army, the U.S. Army, and local citizens combined forces to care for thousands of homeless people at shelters in Cape Girardeau, East Prairie, and Charleston. Others received food and medical attention in rented rooms, with relatives, or in hospitals. Segregated tent cities housed African American refugees.[40] In southern Missouri, as in Mississippi, white and black victims received differential treatment.

High water continued to threaten the Bootheel in the spring of 1937. In early May, emergency workers rushed to Charleston to close the break in the levee. Hundreds of Civilian Conservation Corps and Works Progress Administration workers joined local volunteers to help shore up the levee against a second flood. By May 8, the river had crested at over forty-eight feet, but the levee held.[41]

At this time, Snow experienced a personal tragedy. Lila had been ill for more than a year, and in May she had the last of a series of operations. She died at St. Mary's Infirmary in Cairo, on May 17, 1937, at the age of forty-nine, leaving Snow a widower again, after sixteen years of marriage,[42] with two growing daughters still at home.

Within days of her death, Snow suffered an unexplained paralysis in his legs. Years later, he wrote, "It did not occur to me that I would go down from the blow, but in two days I found that I could not walk because my legs would not hold me, and my feet would not go where they needed to go. After a time I could get about in the house but could not venture from it."[43]

While Snow reeled from his loss, the Bootheel began to rebuild.

39. Ibid., January 28, 1937.

40. *Charleston (Mo.) Enterprise-Courier,* January 28, February 4, and February 11, 1937.

41. *Cape Girardeau Southeast Missourian,* May 4, May 6, May 7, and May 8, 1937.

42. *Charleston (Mo.) Enterprise-Courier,* May 20, 1937.

43. Snow, *From Missouri,* 224.

After the floodwaters had subsided but before the refugees had moved back into the floodway, Snow drove into the area on roads that remained intact. "I saw a picture of desolation that I cannot forget," he wrote in his memoir. In an area less accustomed to disaster, he might have concluded that recovery would take years. "But," he wrote, "I knew that the good earth somehow would be tilled and would be planted to crops in a few weeks' time. It was indeed."[44]

Several months after the flood, Berthe came to him with a long, rambling manuscript, criticizing the Corps and its botched attempt to control the behavior of the Mississippi River. Berthe wanted to publish the tract but lacked funds. Snow, who was bedridden, edited the piece, toned down the rhetoric, and added an epilogue. Together they arranged to have the little book printed under the title *Old Man River Speaks* on slick paper with illustrations, maps, and graphs. The enterprise relieved Snow's depression, but it did not cure his physical weakness. He completed the editing task without leaving his bed.[45]

Snow, Berthe, and local officials struggled with problems in the floodway throughout the 1930s and 1940s. In 1949 county officials, city administrators, and citizens in Mississippi and New Madrid Counties appealed to Congress for a solution to drainage problems exacerbated by the post-1927 flood-control program. Construction of the setback levee had effectively dammed many drainage ditches, leaving much of the farmland west of the levee without an outlet for excess water. The Corps' diversion channel, which was supposed to carry away the runoff, did not function properly because in some places it was as much as eight feet higher than the drainage ditches.[46]

By this time Berthe was gravely ill, but Snow joined the argument, noting frequent flooding in the Big Lake area. He reported to the Corps that sandbags were often required to protect the state road along the lake bank near his property. In order to alleviate this problem he recommended that the Corps improve and extend existing ditches, provide additional ditches, or, possibly, cut a sewer outlet through the levee. He adopted a conciliatory tone, expressing his

44. Ibid., 220.

45. Ibid., 216–17; L. T. Berthe, *Old Man River Speaks*, 34–35

46. Members of the County Courts to Honorable Orville Zimmerman, House of Representatives, [June 14, 1949], Snow Papers, folder 5.

belief that the Corps was "searching just like I am for the right and proper solution to our drainage dilemma."[47]

Clearly, he perceived these floodway troubles as symptomatic of larger social and environmental problems. He was not the same brash pioneer who came to the flood country in 1910. By the end of the 1930s, he had seen four inundations of the Missouri lowlands. The flood of 1912 was a challenge; the second flood, just a year later, was dreary and mean. Before the next disaster, the flood country became ditch-and-levee country, cutover, drained, and walled off from the river. But the floods just got bigger. The overflows of 1927 and 1937 were more than nuisances. They were catastrophes.

Flood control and drainage in the Bootheel were not just engineering problems; they were human problems. Snow believed that there was a direct connection between the 1937 flood and the roadside demonstrations that would take place in 1939. Both events happened in January—in the dreary, damp cold of a Bootheel winter. Both events demonstrated to the wider world the suffering and strength of the farmworkers of southeastern Missouri. As he wrote in *From Missouri*, "I believe there is a relationship between the flood's harsh experience and the comparable hardships that floodway working people deliberately took upon themselves two years later on January 10, when they moved themselves and their small belongings to the roadsides, and exposed themselves shelterless to the midwinter elements."[48]

In 1949, Snow wrote to an officer of the Corps, "I know that people often react emotionally and not logically to our water problems, somewhat like they do in respect to problems of foreign relationships."[49] Beginning in the 1930s, he gradually made connections between all these issues—of floods and levees, wealth and poverty, war and peace.

47. Thad Snow to Colonel L. H. Foote, U.S. District Engineer, West Memphis, Arkansas, August 6, 1949, Snow Papers, folder 5.

48. Snow, *From Missouri*, 223.

49. Thad Snow to Colonel L. H. Foote, undated letter [1949], Snow Papers, folder 5.

King Cotton   4

They had almost turned the country into a northern wheat, corn, and live stock affiliate when an amazing thing happened in 1923. In that year Swampeast Missouri went south almost over night.

Or, better stated, the South came up to envelop and absorb this budding corn belt garden spot, and made it over into a land of cotton plantations, within a year's time.

—Thad Snow, "History of Swampeast Missouri" [1939]

A headline in the *Sikeston Standard* on January 12, 1923, heralded the arrival of "King Cotton in Southeast Missouri." Like Snow, the newspaper used exaggeration for dramatic effect. The cotton economy did not arrive in a single year, but its advent did mark a portentous change in the Bootheel. The newspaper article quoted S. S. Thompson, mayor of Portageville, who boasted that four gins were operating in that town, where the Frisco Railroad loaded fifteen thousand dollars' worth of cotton on one flatcar. Investors in Charleston and Sikeston announced plans to open gins. Thousands of bales of cotton waited in heaps near the Mississippi River docks in Caruthersville for a barge to carry them to New Orleans. Some farmers in the region reported tripling their earnings in 1922 by raising cotton.[1]

In the summer of 1923, Snow still took pride in his corn crop,

---

1. *Sikeston (Mo.) Standard,* January 12, 1923.

asserting that one of his fields contained "the finest patch of corn man's eyes have ever beheld." He continued to raise wheat and alfalfa, but he also planted 150 acres in cotton, confessing that he found the challenge of planting this new crop an "exhilarating experience."[2] Snow keenly observed the Bootheel's plunge into the vortex of the cotton economy.

Describing the influx of new people with new values and goals, he wrote that, "Ginners, cotton buyers, and what not came along. Resident farmers, even those of corn belt background, caught the fever (cotton was high and always would be) and put up thousands of flimsy shacks for the immigrant croppers and changed themselves into cotton planters, on credit, in a few weeks' time." Recruited by planters or lured by the chance to make a new start, according to Snow, "Ten thousand negroes moved up out of the Cotton South."[3] His figure was not exactly correct; the actual number exceeded fifteen thousand. The influx he observed was real.

The transition to a cotton economy and the change in population began earlier and proceeded more gradually than Snow recalled. From the earliest days of settlement, the Bootheel had ties to the South. Many of the first residents had come from southern states, and the region leaned toward the Confederacy during the Civil War. The transition to cotton agriculture reinforced these cultural ties, enmeshing the Bootheel more deeply in the plantation economy that helped to define the South as a distinctive region well into the twentieth century.[4]

Geography and climate predisposed the Bootheel to cotton agriculture. Cotton is a fine, usually white, fiber that grows on the seeds of several species of plants in the genus *Gossypium*. Wild varieties grow in tropical regions in both the eastern and western hemispheres. In the tropics, the plant is perennial, but in the temperate American South, it is an annual. Cotton requires a two-hundred-day growing season without frost and an average summer temperature of at least seventy-seven degrees; it will not grow above a line that roughly corresponds with the border between Arkansas and Missouri and

2. Ibid., June 12, 1923.
3. Thad Snow, "History of Swampeast Missouri," Snow Papers, folder 20.
4. See Irvin J. Wyllie, "Race and Class Conflict on Missouri's Cotton Frontier."

the northern edge of the Bootheel.[5] Only floods, swamps, and accidents of history delayed its inclusion in the American cotton belt.

Planters and farmworkers began surging into the lowlands in the 1880s as the land was cleared and drained. From 1900 to 1920 the rural population of New Madrid, Pemiscot, and Scott Counties more than doubled, while the rural population of Dunklin County increased by fifty percent. In fact, the overall population in all Bootheel counties soared from 1880 to 1930.

Table 1
Population of Bootheel Counties[6]

County	1880	1900	1920	1930
Dunklin	9,604	21,706	32,773	35,799
Mississippi	9,270	11,837	12,860	15,762
New Madrid	7,694	11,280	25,180	30,262
Pemiscot	4,299	12,115	26,634	37,284
Scott	8,587	13,092	23,409	24,913
Stoddard	13,431	24,669	29,755	27,452

While developers cleared and drained the land, the spread of the boll weevil in other parts of the South continually pushed growers west toward this final cotton frontier. Between 1892 and 1922, the destructive insect had eaten away at the crop in every Cotton Belt stronghold, with the exception of northern North Carolina and western Texas. Rupert Vance traced the spread of this scourge:

> In 1903 the weevil had reached the western tip of Louisiana, by 1906 Arkansas. The Mississippi River was crossed in 1907, and by 1910 the weevil had covered southern Mississippi and penetrated into Alabama. . . . After reaching Georgia in 1914, it spread

5. Rupert Vance, *Human Factors in Cotton Culture: A Study in the Social Geography of the American South*, 12–14.

6. Source: U.S. Bureau of the Census, *Fifteenth Census of the United States: 1930*, vol. 3, pt. 1: *Population*, 1341–46, and *Thirteenth Census of the United States: 1910*, vol. 2: *Population*, 1075–85.

rapidly. . . . It touched South Carolina in 1917, swept across the state in two years and virtually covered North Carolina by 1923.[7]

At the height of the epidemic, more than 90 percent of the crop was infested.

By 1920, planters in the Bootheel were turning away from diversified farming to devote more and more of their land to cotton production. There was a dramatic increase in acres planted to cotton in the six counties:

Table 2
Acres Planted in Cotton[8]

County	1909	1929	1934
Dunklin	44,061	89,241	72,854
Mississippi	149	25,239	27,702
New Madrid	9,894	54,240	64,752
Pemiscot	21,688	125,637	94,313
Scott	20	16,182	14,477
Stoddard	8,239	19,490	23,076

In the early 1930s, the Federal Writers Project of the Works Progress Administration reported that cotton had "usurped" all the land in the Bootheel south of the town of Malden in Dunklin County. Nearly all the trees were gone. Corn and wheat production had dwindled. According to the WPA guide to Missouri, "South of Malden, 'King Cotton's' dominion meets the horizon with a monotony that is broken only by occasional clumps of trees in a brush-choked swamp."[9]

7. Vance, *Human Factors,* 95.
8. Source: U.S. Bureau of the Census, *Thirteenth Census Abstract with Supplement for Missouri,* 676–83; and *Fifteenth Census of the United States: 1930,* vol. 2, pt. 1: *Agriculture,* 1017–23, and *Census of Agriculture: 1935, Reports for States with Statistics for Counties and a Summary for the United States,* vol. 1, 280–85.
9. *Missouri: The WPA Guide to the "Show Me" State,* 527.

Geographer Sam T. Bratton confirmed the WPA writers' observation that farmers in the northern Bootheel counties, especially in the Sikeston area, clung to a more midwestern pattern of diversified land use, with some trees remaining on the land. Many farms of two hundred or more acres produced livestock, corn, melons, and berries, but these farms also grew cotton. In the lower Bootheel counties, southern plantation agriculture prevailed. Large landholdings were divided into tracts of ten to one hundred acres and leased to tenant farmers or farmed by sharecroppers. In the southern Bootheel, Bratton observed "endless fields of cotton, plantation homesteads, and grouped cabins of negro tenants."[10]

Between 1900 and 1930, more than seventeen thousand African Americans came to the Bootheel, with more than fifteen hundred arriving between 1920 and 1930. The black population in Pemiscot County grew from 412 in 1890 to 10,040 in 1930, with the number nearly tripling between 1920 and 1930. In Mississippi County, also, the black population tripled in that one decade. Dunklin County's black population remained quite small, but other Bootheel counties saw significant growth.[11]

Table 3
African American Population of Bootheel Counties[12]

County	1900	1910	1920	1930
Dunklin	205	96	147	461
Mississippi	2,265	2,006	1,311	3,997
New Madrid	2,027	2,097	1,950	5,617
Pemiscot	862	1,533	3,865	10,040
Scott	505	545	365	1,531
Stoddard	47	24	17	1,692

10. Sam T. Bratton, "Land Utilization in the St. Francis Basin," 376, 381–82.
11. U.S. Bureau of the Census, *Thirteenth Census of the United States: 1910,* vol. 2: *Population,* 1106, 1114.
12. Source: U.S. Bureau of the Census, *Fifteenth Census of the United States: 1930,* vol. 3, pt. 1: *Population,* 1341–46, and *Thirteenth Census of the United States: 1910,* vol. 2: *Population,* 1106–19.

Most of these immigrants were tenant farmers or sharecroppers, who made up the bulk of the workforce in the cotton economy. They migrated in family groups, not as individual male laborers. Some verged on destitution, but others possessed mules, equipment, or other resources with which to start a new life. In early January 1923, the *Sikeston Standard* reported that "The past few weeks has [sic] brought hundreds of these families into this section and we will state that they are big families, well dressed and every one of them self-sustaining with bank accounts."[13]

Not all tenant farmers and sharecroppers were black. Poor white families also built flimsy houses in cotton fields, paying rent to landlords or working for a share of the crop.[14] In many instances, white farmworkers tried to intimidate the African American families, warning them to leave the area and threatening them with violence. In several towns, including Pascola, Hayti, Steele, Holland, Cooter, and Denton, white residents threw stones or fired shots into black people's houses. Some African Americans left the region, but more arrived in response to offers from white planters.[15]

Jim Mac Emerson of Morley, in Scott County, a cotton farmer and ginner, described the process of recruiting labor for his farm. "Well," he said, "we used to go down in Mississippi. . . . We would go down there on those plantations and get people and bring them up here." The workers he recruited already knew how to plant and pick cotton. "They were what we called shareworkers, sharecroppers. They would raise the cotton for us. We would finance them. They did all the work."[16]

Tomy Lane, a Scott County sharecropper, explained how the system worked. He moved from Mississippi to the Bootheel in the 1930s. The planter furnished the seed, farm equipment, horses, and mules and provided a house for the tenant and his family. During the winter, the planter "would give you so much money, a check,

13. *Sikeston (Mo.) Standard*, January 19, 1923.

14. Acel Price, interview with the author, Van Buren, Mo., September 12, 2002.

15. *Sikeston (Mo.) Standard*, March 3, 1923.

16. Jim Mac Emerson, interview with David Dickey, September 9, 1986, transcript on file in Special Collections and University Archives (Kent Library), Southeast Missouri State University, Cape Girardeau.

every two weeks to feed your family. When you were a boarder, picking cotton, you had that tab going."[17]

In April, the sharecropper planted the cotton using a wooden plough, or seeder, pulled by a mule. The cotton plants bloomed in summer, and bolls formed under the blooms. The blooms turned red, then dried up and fell off the plants. The boll was ready for picking around the Fourth of July. The sharecroppers' whole family went to work then, picking the bolls and putting them in nine-foot-long sacks. Each sharecropping family worked about thirty-five acres.[18]

After that, the planter took the crop to the ginner, who separated the cotton lint (or fibers) from the seeds. In theory, the planter and the tenant shared the crop fifty-fifty. As Tomy Lane explained, however, "When you go to picking cotton, well, you see, he [the planter] takes his part out of there—what you owe him. Just like you went and borrowed money from him. Just like those checks that he let you have to live off until the cotton is made. He takes that out of your half."[19]

Alex Cooper, son of an African American farmer in the Bootheel, described the social hierarchy of the cotton lands. At the top of the pyramid were the landholders, who owned large and small farms. Next in status were the leaser-farmers (renters or tenant farmers), who paid cash rents and owned their own mules and equipment. Beneath them in the hierarchy were sharecroppers, who worked for half the crop. Lowest on the social scale were day laborers, who had salable skills such as mule breaking or cotton picking.[20] Day laborers sometimes worked for sharecroppers, who paid them seventy-five cents to a dollar and a half per day, plus board.[21]

In the twentieth century, sharecropping was an anachronism, based on an outmoded system of paternalism. When cotton came to the Bootheel, merchants, professionals, corporations, and other investors bought large tracts of land, operating plantations as absen-

17. Tomy Lane, interview with David Dickey, August 13, 1986.

18. Ibid.

19. Ibid.

20. Alex Cooper, interview with David Whitman, March 11, 1994, Portageville, Mo., Bootheel Project, audiocassette 15, Western Historical Manuscript Collection, University of Missouri–Columbia.

21. Price interview, September 12, 2002.

tee landlords.[22] As plantations became businesses, the tradition of paternalism survived in vestigial form while personal bonds between planters and farmworkers eroded. As time went on, planters depended more and more upon day laborers, who received cash wages, or "scrip," and had no security from season to season.[23] A continuing influx of farmworkers into the area created a labor surplus, turning the Bootheel into what one scholar described as a "rural slum."[24]

Like other planters, Snow took advantage of the farm laborers who crossed the state line, often in broken-down wagons, looking for work. He built two-room wooden houses for them or gave them the boards and nails to build their own houses. For the first six months of the year, before picking time, he furnished them with food and essentials. Then, at picking time, he took his share. During the 1920s and 1930s, he housed as many as twenty-three black families at Snow's Corner. In his memoirs, he commented drily on the advantages of the outworn labor system—for landowners. "We enjoyed a new and unaccustomed position of overlordship that was mildly intoxicating. It is nice, as everybody knows, to have employees and to direct their work, but it is much nicer to have servants who don't talk back."[25]

The intoxication was short-lived. In 1929, the stock market crash sent the cotton economy into a tailspin. With the textile industry in crisis, prices for raw cotton fell drastically in the early 1930s. According to Alex Cooper, the Great Depression pushed sharecroppers to the bottom of the social system. When landlords evicted them, they had nowhere to go. Croppers who had recently migrated to Missouri could not go back to the southern states from which they came.[26]

---

22. Vance, *Human Factors*, 63-65.
23. Alex Lichtenstein, "The Southern Tenant Farmers' Union: A Movement for Social Emancipation," introduction to Howard Kester, *Revolt among the Sharecroppers*, 27–28.
24. Carey McWilliams, *Ill Fares the Land: Migrants and Migratory Labor in the United States*, 285.
25. Snow, *From Missouri*, 159, 155. It has not been possible to identify these families in the 1930 U.S. Census; sharecroppers are not listed under the household of the landowner.
26. Dewey W. Grantham, *The South in Modern America*, 116–17; Alex Cooper interview, March 11, 1994.

In southeastern Missouri, Snow wrote, "The first three years of the thirties were a nightmare." Foreclosures became commonplace. Many farmers simply turned over their mules, equipment, and property to creditors. Others endured the humiliation of arranging all their belongings for a sheriff's sale outside the courthouse in Charleston. When Snow lost some acreage to foreclosure and had to seek help from his in-laws to keep his farm, he did not feel lonely; he recalled that "Farmers went down like ten pins or a row of dominoes. I don't mean that absolutely all farmers or all town people went broke, but in my county so few remained solvent that they might have been counted on the fingers of both hands."[27]

Like most southern farmers, Snow welcomed the New Deal with open arms. When President Franklin D. Roosevelt took office in 1933, the price of cotton had dropped to six cents a pound, one third of its 1929 price. Convinced that overproduction had caused the drop in prices, the Roosevelt administration proposed federal legislation to reduce the size of the cotton harvest.[28]

In May 1933, when Congress passed the Agricultural Adjustment Act (AAA), the cotton crop was already in the ground. Under this program, the government offered cash benefits to farmers who would take portions of their land out of production. In 1933, the government actually paid farm owners to plow up a portion of their cotton. According to Snow, Bootheel planters complied with Washington's request to plow up a third of their acreage.[29]

While they acquiesced in crop reduction plans, planters complained about the administration of the AAA programs. In 1934, there was a great deal of bickering about faulty statistics on percentages and yields. New Deal bureaucrats apparently miscalculated the productivity of the lowlands, thereby reducing subsidies. Planters wrote irate letters to Secretary of Agriculture Henry Wallace and his special assistant, Julien Friant, a citizen of the Bootheel.[30] But 1934 was a good crop year, and planters began to pull out of the depths of the Depression. When more than four thousand farmers traveled to

27. Snow, *From Missouri*, 169, 179.
28. Lichtenstein, introduction to Howard Kester, *Revolt among the Share-croppers*, 30.
29. Grantham, *South in Modern America*, 157; Snow, *From Missouri*, 180.
30. Letters on file in the Friant Papers and the Caverno Papers.

Washington in 1935 to thank Roosevelt and Wallace, Snow was in the forefront of the crowd.[31]

In general, New Deal farm recovery programs helped planters but exacerbated the troubles of tenant farmers and sharecroppers. Many large farmers plowed up their land and evicted their tenants. The law required landlords to share subsidy payments with tenants, but many landlords kept the benefits and ignored the law. The socialist leader Norman Thomas observed that the operation of the AAA program threw thousands of farm laborers off the land with no hope of employment in industry or agriculture.[32]

Following the example of the planters, southern agricultural laborers turned to the federal government for help. Washington had already provided disaster assistance during the flood of 1927 and the drought of 1930–1931. In response to rising unemployment and homelessness in the rural South, the Federal Emergency Relief Administration (FERA) and the Division of Subsistence Homesteads in the Department of the Interior constructed more than one hundred planned agricultural communities, mostly in cotton-producing states.[33]

After the United States Supreme Court struck down key elements of the AAA programs, other agencies continued to promote agricultural recovery, with mixed results. Between 1935 and 1943, the Resettlement Administration and the Farm Security Administration (FSA) of the United States Department of Agriculture made loans to nearly four hundred thousand southern farm families, purchased thousands of acres of nonproductive farmland, established cooperative agricultural communities, and operated camps for migrant workers. Despite these efforts, historian Dewey Grantham concluded that New Deal programs "made hardly a dent in relieving the problem of the depressed landless farmer" in the South.[34]

New Deal officials investigated conditions in southeastern Missouri in 1936 and 1937. The FSA, established in 1937 within the USDA, published a shocking report on the Bootheel under the title *Rich Land—Poor People* in January 1938. In the chapter "How the People

---

31. Webster Schott, "Thad Snow, the Farmer," 17.
32. Grantham, *South in Modern America,* 157.
33. Ibid., 158.
34. Ibid., 160.

Live," the authors stated that sharecroppers and farm laborers made up the majority of the population but had little influence on economic and social conditions. Their incomes were "too low to maintain a decent standard of living," even with benefit checks and relief from the government.[35]

Housing conditions, according to the report, were "so far inferior to other parts of the State as to make extremely difficult any comparable measurement." A typical dwelling was the strip house that had walls made of vertical boards and wooden strips over the cracks to keep out the weather. More than half of the white families and nearly two-thirds of the black families lived in this kind of house. Only a third of the white families and 15 percent of the black families could afford weatherboard houses—frame houses with horizontal siding. Because of the wet soil, most houses had no cellars but were set on concrete blocks or slabs or wooden piers. Dogs and chickens found shelter in the crawl spaces under the houses. Most of the farm laborers' houses had no plaster, insulation, or wallpaper.[36]

Poor housing and sanitation caused health problems of tragic proportions. Mortality rates for preventable diseases were extremely high in the southeastern lowlands. Two-thirds of all Missouri's deaths from malaria in the 1930s occurred in the seven Bootheel counties, where swamps and drainage ditches bred mosquitoes and many houses lacked adequate screening. Deaths from typhoid fever declined in Missouri between 1926 and 1934, but the rate in the Bootheel remained much higher than in the rest of the state. The region also suffered from a high incidence of pneumonia and gastrointestinal diseases. The infant mortality rate was much higher than in any other area of the state.[37]

The Bootheel's farmworkers were a transient population. Throughout the South, tenants moved often from farm to farm. As New Deal investigators observed, "With little or nothing to lose, the tenant moves constantly with the hope of improving his lot." The situation in Southeast Missouri was even more unstable. According to the 1938 report:

35. Max R. White and Douglas Ensminger, *Rich Land—Poor People*, 37–38.
36. Ibid., 40–42.
37. Ibid., 47–51.

In Southeast Missouri, this mobility is accelerated. Laborers are imported from Arkansas and Tennessee during the cotton season and return at the end of the season. Levee workers move in and out of agriculture. The ownership of land changes and a shake-up in the tenants occurs. The land is flooded and the tenants move on, their places being taken by tenants who move in.[38]

This constant movement made it difficult or impossible for state authorities to enforce school attendance laws or administer government relief programs in Bootheel counties, which swelled from 20 to 50 percent in population between 1930 and 1940.[39]

As the size of landholdings increased, while the number of farms decreased, many sharecroppers could no longer find a piece of land on which to make a living. The FSA estimated that between 1926 and 1936 more than 60 percent of the sharecroppers in Southeast Missouri had to look for jobs as day laborers. For these workers employment was seasonal, and they no longer had shelter—not even a shack—or supplies to tide them through the winter. Most turned to government relief, looked for jobs picking crops other than cotton, or headed west.[40]

In an attempt to provide subsistence for dispossessed workers, the FSA established resettlement communities, or colonies, in the Bootheel. These colonies were segregated, with separate sections for whites and blacks. After the flood of 1937, the FSA created Southeast Missouri Farms, a haven for sharecroppers on sixty-seven hundred acres near La Forge in New Madrid County. Approximately one hundred families—sixty white and forty black—took part in the experiment, under the guidance of Hans Baasch, a Danish-born New Dealer.[41]

With the help of FSA engineers, the men of La Forge built two-bedroom wooden houses and spruced them up with white paint. There was no running water, but each house had a privy, a well, and a pump. The houses also had screened porches, living rooms,

---

38. Ibid., 53.

39. Ibid., 54; U.S. Farm Security Administration, "Southeast Missouri: A Laboratory for the Cotton South," December 30, 1940, p. 1.

40. U.S. Farm Security Administration, "Southeast Missouri: A Laboratory," 2.

41. Stuart Chase, "From the Lower Depths."

kitchens with wood- or coal-burning stoves, built-in cabinets, and enameled sinks.[42]

In the La Forge colony, farmers worked the land and pooled their profits. The government loaned money at low interest so that farmers could purchase mules, supplies, and equipment. Cotton remained important as a cash crop, but the members of the cooperative also raised cattle, hay, chickens, hogs, and vegetables. The cooperative also established a library, a night school, knitting clubs, softball clubs, and churches.[43]

Of course La Forge was not free of problems. Some families borrowed government money and spent it foolishly. A failed cotton crop could destroy the cooperative's economic base. Racial antagonism existed and could flare up at any time. The program was costly and vulnerable to political changes in Washington.[44] Most importantly, resettlement colonies like La Forge helped only a minority of the distressed farm laborers in the cotton belt.

Snow approved of the resettlement projects and criticized those who tried to block them. In May 1937, he castigated his white neighbors for rejecting such a project. He noted that local citizens seemed willing to offer financial inducements to bring a shoe factory into the county, but the same citizens balked at allowing the government to find a resettlement colony for black farmers. He explained the racial and class ideology that prompted opposition to government aid for African Americans: "There are, of course, deep-lying reasons why we are against the proposed Negro colony. We are a cotton country and Negroes do our work. They do it cheaply, unless they catch us in a pinch at cotton-picking time. They know their place and keep to it pretty well in the daytime." Living together in a colony, the black farmworkers would have time to socialize and to share ideas. "Who knows what fool ideas might be generated in a 'resettlement' and overflow the whole countryside?"[45]

For Snow, the Depression gave rise to a great storm of "fool ideas" that he hoped might ultimately lead to solutions for society's most troubling problems. In late 1931, he wrote that people tended not to

42. Ibid., 110.
43. Ibid. 111.
44. Ibid., 112.
45. Thad Snow, "Mississippi County's State of Suspense," *St. Louis Post-Dispatch* [May 1937], clipping in scrapbook, Snow Papers.

think very much "except under pressure of necessity." The Great Depression put such pressure on them, "for one has only to open a conversation with the first hitchhiker on the road to be convinced that thinking has become a national habit." For Snow, the compelling subject of thought was "the amazing paradox of our time—appalling want amid abundant plenty."[46]

In the early years of the crisis, he embraced the New Deal and tried to remain optimistic. By 1934, however, he came to the conclusion that the system of crop reduction through the quota system did not work well in the Bootheel. On October 20 of that year, he wrote a long letter to Victor E. Anderson, Special Attorney for the AAA, in Washington, describing the problem in Mississippi County. The cotton crop in the county was much larger than the government calculated. Under the law, farmers had to pay taxes on the portion of their crop that exceeded a certain quota. In general the landowners and renters still benefited from the New Deal program. But the sharecroppers were, in his words, "most injured and more deeply resentful."[47]

He placed the blame partly on flawed programs and partly on the duplicity of the white planter class. Some planters, he asserted, swindled the government by falsely claiming to practice soil conservation or crop reduction. Others cheated the renters and sharecroppers by failing to pay them their share of government subsidies. Many freely admitted these transgressions, seemingly feeling no shame or social stigma for "Chiseling off the Government."[48] The result of all the bungling and cheating was unconscionable human suffering. Snow observed this resentment among the workers on his own farm.

After Snow signed his crop-reduction contract with the government, he called his workers together for a talk. He explained to them that they would have to reduce their production to comply with his contract. This meant that each one would have to cut production by nearly one half. But he assured them that they would receive a parity payment and that they would not have to pay any tax on their crop. They might have chosen to leave and work for a noncooperating farmer, who would not require them to reduce their

46. Snow, "When Traders Rule," 113.
47. Thad Snow to Victor E. Anderson, October 20, 1934, carbon copy in Friant Papers, copy in Snow Papers.
48. Ibid.

crop, but then they would receive no parity payment and might have to pay a tax. According to Snow they all decided to stay.[49]

Unintentionally, Snow had made false promises to his sharecroppers. He spoke to his workers before the government actually set production quotas. As it turned out, he was allowed, or allotted, 380 pounds of cotton per acre. After halving their output, his sharecroppers produced between 480 and 580 pounds per acre. After reducing their crop further to avoid tax payments, his sharecroppers still had to pay tax on more than a third of their crop.[50]

The sharecroppers resented what they considered unfair treatment. Snow wrote to Anderson: "I am sure that even now none of them believes that I knowingly misled him last spring. But of course they all know that I didn't know what I was talking about, and they labor under a sense of injustice and betrayal." He felt embarrassed by the role he unwittingly played in duping the sharecroppers, and he empathized with their frustration. As a landowner who attempted to deal fairly with laborers, he deeply regretted the situation and felt personally betrayed by governmental bungling. At the close of his letter, he stated, "I could very easily become emotional writing about the wrongs of the sharecropper."[51]

As early as the fall of 1934, sharecroppers in Mississippi County were mobilizing against the injustices of New Deal cotton policy. On November 4, 1934, Snow wrote to his friend Friant in the USDA. In that letter, Snow argued that the system was driving sharecroppers from the land and might push them to take desparate measures.[52] He said he had been asked to address a large crowd of African Americans in Charleston, but he reported, "I just funked the job." He was embarrassed to talk to them. He wanted to defend the AAA and believed that the problems that existed could be solved. But time was of the essence. "We need quick action," he said, "Because some of the poor devil share-croppers get away and for other reasons delay makes it hard to do even justice."[53]

Three years later, Snow saw little improvement in the situation and continued to worry that injustice would lead to rebellion and

49. Ibid.
50. Ibid.
51. Ibid.
52. Thad Snow to Julien Friant, November 4, 1934, Friant Papers.
53. Ibid.

violence. In 1937, he wrote a letter to Governor Lloyd C. Stark, enclosing a clipping from the *Post-Dispatch*. In the letter, dated August 19, he urged the governor to read the article from the previous Monday's paper and then write "short notes to our attorney Jim Haw and sheriff Walter Beck" in Charleston. The clipping, missing from Stark's files, in all likelihood contained a story about the killings of a miner and a sheriff in Harlan County, Kentucky, as a result of labor unrest. Snow asked the governor to praise officials in Mississippi County "because you note that there has been no act of terrorism here and express confidence that they will guard civil liberties faithfully." Snow feared that in tense economic times, Missourians might "go to beating and killing these poor devils for trying to improve their lowly estate by organized effort."[54]

On January 2, 1938, Snow wrote to Charles G. Ross of the *St. Louis Post-Dispatch* that conditions in the Bootheel were horrible and getting worse. He said he had recently arrived home from a trip and had spent most of the day "talking with negroes who have been coming to the back door to see if I couldn't 'do something.' "[55] Some came from neighboring farms, and some came from twenty miles away.

From personal observation, Snow reported that many sharecroppers, both black and white, subsisted on one meal a day of cornmeal, water, and salt. Government relief officials had nothing to offer them. The sharecroppers were slowly starving and were "supposed to do so silently, while hoping for an early spring to provide greens which are wholesome if unsatisfying."[56]

In his letter to Ross he warned that farmworkers might justifiably take dramatic action against injustice, reporting that "Several hundred negroes are gathered together in a certain school house at this moment (8:30 p.m.) debating the advisability of staging a preliminary semi-riot or demonstration before they get too weak to make it seem realistic. The leadership is pretty level-headed, so I doubt if there will be any disorder at this time." He urged Ross to send reporters to come to the Bootheel and speak with relief officials,

54. Thad Snow to Governor Lloyd C. Stark, August 19, 1937, Stark Papers, folder 6973; *St. Louis Post-Dispatch*, August 16, 1937.

55. Thad Snow to Charles G. Ross, January 2, 1938, Snow Papers, folder 1.

56. Ibid.

sharecroppers, and a planter or two. In Mississippi County, he advised, "We are naive enough to speak right out. Down in Pemiscot or Dunklin the planters have learned to be guarded in statements."[57]

Most planters, according to Snow, were oblivious to the threat of a sharecroppers' revolt. In the closing paragraph of his letter, he stated enigmatically that "All is not lost however, because the fortitude and enterprise of the planters are entirely unaffected by the belly pains of the croppers."[58] This "fortitude and enterprise" had enabled the planters to escape the boll weevil and establish a thriving cotton economy in the Bootheel. Possibly, Snow implied, these qualities would help them to bluster their way through the Great Depression and avoid a social revolution. They had lived through other disasters, including a series of floods, and they might live through this one. But their resilience seemed to rest upon a terrible blindness. They were strong and successful, in part at least, because they paid no heed to the suffering and anger of their workers.

Between 1934 and 1938, Snow repeatedly expressed the idea that the federal government had to take action to curb abuses in the system. He had not given up on the idea of farm subsidies and production control, but he believed that the government had not even begun to "undertake seriously the difficult task of cleaning up the thievery and skullduggery that have thrived amazingly under the farm programs in many localities." In Missouri's cotton country, he admitted, the problem was worse than in many other places. "Everybody knows that outlawry on the border is strictly within the American tradition." It was, he insisted, time for the cotton frontier to become "civilized."[59]

Throughout this period, Snow continued farming with sharecroppers. He did not turn to the use of day laborers, as many farmers did. If anything, as Alex Cooper explained, day laborers would have been even more vulnerable to the perils of the time than sharecroppers. Hunter Rafferty recalled that "[Snow] kept his crew of sharecroppers pretty constantly on his place." During the roadside demonstrations, according to Rafferty, "I don't remember any of

57. Ibid.
58. Ibid.
59. Thad Snow, "Farm Control: An Inventory," *St. Louis Post-Dispatch* [1938], clipping in scrapbook, Snow Papers.

them moving out on the highways."[60] However, at least two of Snow's farm workers, a man named Jake Reed and a woman named Annie, would join the roadside protest in 1939.[61]

When Snow succumbed to the "cotton fever" in the 1920s, he could not have foreseen the dramatic events of the 1930s. Like most of his neighbors, he initially acted from self-interest, expecting to make a profit from land well-suited to cotton production, using sharecroppers' labor. By 1934, his letters to the federal government revealed a conviction that he and his landowning neighbors had lost control of the situation; by that year he was speaking and listening to his farmworkers. Whether or not he fully understood their situation, he was at least attempting to communicate with the sharecroppers, especially the African American sharecroppers, and he heard them talking about social revolution.

When the Bootheel joined the cotton belt, the region inevitably became entangled in the web of southern history. By the late 1930s, Snow believed that history had reached a great turning point and that angry sharecroppers might lead the way to a radically different future.

60. Alex Cooper interview, March 11, 1994; Rafferty interview, May 28, 2002.
61. H. G. Simpson to Governor Lloyd C. Stark, January 21, 1939, Stark Papers, folder 1959.

Out On
Mr. Snow's Farm 5

We gonna roll, we gonna roll, we gonna roll the union on.

We gonna roll, we gonna roll, we gonna roll the union on.

—John L. Handcox

Snow correctly sensed the possibility of revolution among the sharecroppers. In the early months of 1934, the Communist Party of the United States of America (CPUSA) sponsored a three-week training school in St. Louis for party organizers, both white and black, from the cotton belt. Ten individuals, including share-croppers and tenant farmers from several states, attended classes on the fundamentals of communism, race problems in America, and organizational problems of the newly created Share Croppers' Union (SCU). "Comrade M." of Alabama taught the classes on labor organizing. To protect them from persecution, the five white and five black students assumed false names and addresses in St. Louis. It seems likely that the instructor, Comrade M., was Marcus ("Al") Murphy, who would later move from Alabama to Missouri.[1]

After the school ended, Murphy continued organizing farm laborers in Alabama, requesting one thousand copies of party mem-

1. "Report on Session of Farm School for Sharecroppers," April 1934, Communist Party of the United States of America Records, microfilm edition, reel 287, folder 3714, Library of Congress Manuscript Division (hereinafter cited as CPUSA Records).

bership books in April 1934.[2] The "Farm School on Wheels" made a circuit around the country, training organizers and recruiting members. Party leaflets specifically targeted sharecroppers, proclaiming that the New Deal benefited only the rich while "Millions of tenants and sharecroppers are unable to live like human beings. They are straggling along highways, frequently evicted without notice, starving and denied relief."[3] Murphy and other party organizers reported that the message appealed to the sharecroppers, who were anxious to know what was going on with farmworkers throughout the nation.

Despite some allegations in the 1940s, Snow was not a communist and never became affiliated with the CPUSA,[4] although, beginning about 1934, he held the view, common to the party, that the New Deal did not treat farmworkers fairly. He understood that if sharecroppers continued to suffer, they would resort to drastic and possibly violent action. In the mid-1930s, he violated the unwritten code of behavior for men of his race and class by encouraging his sharecroppers to join a labor union. He did not promote the SCU, however; instead, he helped an Arkansas-based organization, with socialist ties and an interracial membership, called the Southern Tenant Farmers' Union (STFU).

The STFU originated in a small schoolhouse in Tyronza (Poinsett County), in the delta land of northeastern Arkansas, in July 1934. Howard Kester, one of the union's founders, described its original members as "black and white men, clad in overalls."[5] At that first meeting, the main topic of discussion was how the sharecroppers could secure their share of benefits under the AAA program. As individuals, the workers had no power to make the planters treat them fairly. They needed a union to stand up for them.

---

2. Letter from "Puro" to Comrade Harry, April 16, 1934, reel 287, folder 3714, CPUSA Records.

3. "Farmers and Sharecroppers Demand an End to Poverty and Misery" (1935 leaflet), reel 287, folder 3714, CPUSA Records.

4. Fred J. Romanski, archivist, Civilian Records, Textual Archives Division, of the National Archives, discovered a file among the General Records of the Department of Justice (Record Group 60), containing a reference dated March 13, 1947, to security matters on the part of Thad Snow as an alleged member of the Communist Party. However, the Department of Justice File (146–1–42–60) containing that reference was destroyed in 1991 pursuant to a federal records disposal schedule. There is no evidence that Snow belonged to the Communist Party at any time.

5. Kester, *Revolt among the Sharecroppers*, 55.

Shortly after the first meeting, the group recruited H. L. Mitchell, who owned a small dry-cleaning business in Tyronza, and Clay East, who ran the filling station next door to him. Mitchell had once been a sharecropper. He and East had a reputation among local planters as socialists, or Reds, who constantly came up with strange ideas about politics, economics, and labor. Mitchell and East signed on as organizers and made nightly trips to churches and school-houses in northeastern Arkansas, spreading the union gospel.[6]

Formally incorporated in Arkansas on July 26, 1934, the STFU had a strong base in rural Christianity. Many of the union's leaders were lay ministers and backwoods preachers. Kester grew up in rural Virginia and trained for the ministry, although he was never ordained. Sharecroppers avoided the mainstream churches but clung to strong Christian beliefs. STFU organizers won the trust of black and white farm workers with religious songs and imagery. Handcox, the union's troubadour, often used religious verse in his songs. Nearly half of the union's fourteen executive council members in 1935 were preachers.[7]

In Missouri, the STFU recruited Owen Whitfield, who used his sermons to bring the Bootheel sharecroppers into the union fold. Whitfield was born in Jamestown, Mississippi, in 1894, the son of sharecroppers and the descendant of slaves.[8] He and his young wife, Zella, migrated to the Bootheel during the cotton boom, just before Zella gave birth to her sixth child. She would eventually raise seven daughters and five sons.[9] Owen, Zella, and the children all worked in the cotton fields. On Saturdays, Zella took the day off to do the family's laundry.

By the mid-1930s, Whitfield had begun preaching. He continued to work in the fields six days a week, and on Sundays he acted as pastor to other sharecropper families. Two of his children had died, but others were born. He, Zella, and the children slept four to a bed, two at each end. Gradually, his older sons took over the farmwork, and he traveled around the Bootheel, preaching about social justice and the union.[10]

6. Ibid., 58.

7. Lichtenstein, introduction to Howard Kester, *Revolt among the Share-croppers*, 18.

8. Cadle, "Cropperville," 16.

9. Farmer interview, April 7, 2002.

10. Cedric Belfrage, "Cotton-Patch Moses," 96–97.

In addition to Christianity, the STFU had strong ties to the rural socialism of the early twentieth century. This grassroots radicalism was particularly strong in Arkansas, Louisiana, Oklahoma, Texas, and the Bootheel.[11] Socialist newspapers like the *Girard (Kans.) Appeal to Reason,* the *St. Louis Rip-Saw,* and the *Scott County (Mo.) Kicker* inspired class-conscious analyses of social problems in these rural areas. Orators like Kate Richards O'Hare rallied farmers at open-air meetings throughout the South and Midwest. Presidential campaigns by Eugene V. Debs in 1912 and 1920 won an impressive number of votes in the Mississippi Delta region. Socialist leaders like Norman Thomas shaped the views of Kester and Mitchell, who became dominant forces in the STFU.[12]

In December 1934, Mitchell and four other representatives of the union got in an old car and drove from Memphis to Washington, D.C., and met with Secretary of Agriculture Henry A. Wallace. The delegation of three white men and two black men reported that when sharecroppers complained to federal officials about unfair treatment, the officials referred the matter to the planters, who abused the system established by the AAA program.[13] The program already had its problems, with serious internal ideological divisions. But Wallace gave the STFU representatives a sympathetic hearing and promised to investigate the situation.

After the meeting in Washington, Mitchell asked a Methodist minister named Ward Rodgers to call a meeting of STFU members. In mid-January 1935, Rodgers addressed a crowd of sharecroppers at Marked Tree in Poinsett County, Arkansas. A group of planters had previously accused him of teaching Marxist doctrine to the farmworkers. As he addressed the crowd, he grew heated, venting his anger at the planters and even threatening violence against them. Mitchell tried to calm him down, but the crowd responded wildly. Eventually, the delegation that had just returned from Washington, D.C., took the stage and reported on their meetings with federal officials.[14]

11. Ibid., 23. For a discussion of socialism in the Bootheel, see Leon Parker Ogilvie, "Populism and Socialism in the Southeast Missouri Lowlands."

12. Lichtenstein, introduction to Howard Kester, *Revolt among the Sharecroppers,* 25.

13. H. L. Mitchell, *Mean Things Happening in This Land,* 56–57.

14. David Eugene Conrad, *The Forgotten Farmers: The Story of Sharecroppers in the New Deal,* 155–57.

The meeting ended peacefully, but as the crowd dispersed, the Poinsett County Attorney arrested Rodgers for attempting to overthrow the government of Arkansas. His trial and conviction became a national spectacle. He was sentenced to six months in jail and a five-hundred-dollar fine, but he was freed on bond pending an appeal that was never heard. He never served his sentence.[15]

After Rodgers's trial and conviction, planters in northeastern Arkansas repeatedly terrorized members of the union. Local authorities arrested union leaders. Vigilante groups padlocked churches and packed schoolhouses with hay to keep the union from holding meetings in them. Tenants who joined the union received eviction notices. Other farmworkers reported floggings and beatings. Both planters and tenant farmers began carrying weapons. Despite the threat of violence, hundreds of sharecroppers, along with their wives and children, flocked to open-air meetings. Planters, bosses, sheriffs, and deputies would shoot over the heads of the crowd to try and disperse the unionists.[16]

In March 1935, Norman Thomas toured Arkansas and spoke to thousands of sharecroppers. On the final day of his journey, he planned to give an address in a black church in the town of Birdsong. When Kester stood to make the introductions, armed hecklers jerked him from the stage. Thomas waved a copy of the state constitution and insisted that the meeting was legal. Brandishing guns, a group of inebriated men came up behind him and pushed him from the platform. The sheriff advised him to leave town before somebody got hurt. A caravan of automobiles escorted him to the county line. Speaking on national radio, Thomas reported that there was "a reign of terror in the cotton country of eastern Arkansas" and predicted that it would end in the complete suppression of the union or in a bloodbath.[17]

The STFU managed to survive, despite the charged atmosphere in Arkansas. Union members remained staunchly nonviolent and clung to the Christian underpinnings of their movement. The interracial nature of the union defused some of the racial antagonism that could have led to greater violence. An exclusively African American union might have been doomed. The presence of white sharecroppers

15. Ibid., 157–58.
16. Kester, *Revolt among the Sharecroppers,* 60–63.
17. Ibid., 85.

at meetings probably quelled brutal impulses. Women and children also participated in meetings and demonstrations, and this probably had a calming influence on hecklers and foes. Racial and class feelings ran high, but the Great Depression had an impact on people's attitudes toward workers and the poor. The STFU emerged at a time when workers of all kinds flocked to the unions.

The Roosevelt administration had legitimized unionism. In 1935, the National Labor Relations Act (Wagner Act) asserted the workers' right to organize and bargain collectively. Prior to that time the American Federation of Labor (AFL) promoted craft unions for skilled workers, who were mostly white males. In the mid-1930s, the Congress of Industrial Organizations (CIO) challenged the AFL, organizing workers on the basis of industry rather than craft. The Wagner Act, Roosevelt's public support of unionism, and the aggressive spirit of the CIO leader John L. Lewis inspired previously unorganized workers, including blacks, immigrants, and women, to join locals.

In the spring and summer of 1936, the STFU went out on strike in northeastern Arkansas. By May 18, union members began marches through the countryside to call attention to the anger of the workers. The governor of Arkansas called out the National Guard.[18] Mitchell, the secretary of the STFU, reported on May 20 that twenty-five hundred workers, representing seventy-six locals in Cross, Crittenden, and St. Francis Counties, had walked off the job. The strike continued for about a month, with constant threats of violence.[19] A planter in Cross County who agreed to bargain with his workers suffered ostracism. The bank threatened to close on his mortgage. Ginners refused to accept his cotton. Under this pressure, the planter evicted his tenants.[20]

John Handcox, an Arkansas hoe man, found himself jobless during this violent season. He had expressed the sentiments of the STFU ever since he appeared at the union office in Memphis in 1935 with this four-line poem, written in pencil on a piece of tablet paper: "When a sharecropper dies / he is buried in a box / Without any

18. Schroeder and Lance, "John L. Handcox," 127.
19. *Cape Girardeau Southeast Missourian,* May 20, May 29, June 1, June 16, and June 17, 1936.
20. Lichtenstein, introduction to Howard Kester, *Revolt among the Share-croppers,* 52.

necktie / and without any sox."[21] Mitchell recruited him as an organizer and began printing his songs in *The Sharecroppers' Voice*. In
1936, he witnessed the violence in St. Francis County. During this
time, he coined the phrase "There is mean things happening in this
land," which later became the title of Mitchell's account of the
struggles of the STFU.[22]

On March 9, 1937, Handcox visited the Library of Congress and
recorded several of his compositions for the Archive of Folksong. At
that time, he was on a fund-raising tour for the Socialist Party. But
years later he told an interviewer that, at the time of this recording
session, he was living on Snow's plantation.[23] According to Mitchell,
Snow had invited the union to the Bootheel. As Mitchell recalled,
"Thad Snow, who owned a large cotton plantation in southeast
Missouri, actually invited the union to send an organizer. John L.
Handcox, who was chosen to go, afterward sent me [a] poem."[24]

In this poem, written in 1936 or 1937 and entitled "Out on Mr.
Snow's Farm, or The Kind of Man We Like to Meet," Handcox called
Snow "one of the best men in SE-MO." The poet sang about going
to Snow's Corner and receiving permission to organize the workers
on Snow's farm. At a time when other planters were terrorizing
union members, Snow's behavior was truly extraordinary, and
Handcox "rejoiced," saying, "This is the kind of men we need you
know / Men that are in sympathy with their labor, like Mr. Snow."[25]

The song related a detailed account of Handcox's initial meeting
with Snow. As Handcox recalled it:

> Early the second Monday in June
> I walked up to Mr. Thad Snow's Home, all alone
> And introduced myself to Mr. Snow
> One of the best men in SE-MO I know.

This was probably the second Monday in June of 1936, during the
strike in northeastern Arkansas.

During this meeting, Snow and Handcox had a conversation

21. Schroeder and Lance, "John L. Handcox," 123.
22. Ibid., 127.
23. Ibid., 123.
24. Mitchell, *Mean Things Happening*, 349.
25. The poem is printed on page 349 of Mitchell's book.

about the sharecroppers' troubles—a very unlikely conversation between a union organizer and a planter. Even more surprisingly, Snow
encouraged Handcox to speak to his croppers. Handcox's poem
continued:

> He says to me, something for labor ought to be done
> And you are perfectly welcome to go on my farm,
> He pointed me out some of the hands in the field,
> Told me to talk with them and see how they feel.
> Then he asked me what else he could do
> To help put our labor movement through.
> I told him his help would be much if he didn't object
> For the labor on his farm to join our Union as such.

Snow did not object, and Handcox proceeded to recruit the workers at Snow's Corner for the union. The planter's unusual behavior
filled Handcox with optimism:

> I walked over Mr. Snow's farm in all ease
> For I knew that Mr. Snow was well pleased.
> I sang. I talked and rejoiced as I went,
> For I knew I had gotten Mr. Snow's consent.
> This is the kind of men we need you know
> Men that are in sympathy with their labor, like Mr. Snow.

Mitchell later wrote that Handcox asked him to send a copy of
the song to Snow. Since no copy of the verse survived in the Snow
papers, it is possible that he never received the poem. The poet later
recalled that he had very little contact with Snow after their first
meeting. Nevertheless, Handcox felt welcome on the plantation and
stayed with a family there for several months and perhaps returned
there the following year.[26]

In his memoirs, Snow did not describe the meeting memorialized
in the song. However, he did state that he met Owen Whitfield in
the spring of 1936 and became involved in the pickers' strike.[27] Whitfield may have prompted Snow to contact the union leadership in

26. Schroeder and Lance, "John L. Handcox," 126.
27. Snow, *From Missouri,* 235.

Memphis. It seems likely that Whitfield, who was a union leader, assured his colleagues that an organizer such as Handcox might expect Snow to receive him cordially.

By the spring of 1937, officers of the STFU in Memphis were well aware of Snow's support for organizing efforts in Missouri. On April 15, 1937, William R. Anderson of the University of Tennessee in Memphis wrote to Snow to commend him for his courage. Anderson stated that the STFU office in Memphis had informed him that Snow had taken the lead in aiding a local in the Bootheel. "I can only hope," said Anderson, "That your example will soon be followed by other landlords."[28]

In 1937, the CIO set about creating a new organization, for farm-workers. By this time, the STFU represented thirty thousand agricultural workers in seven states. Despite this fact, John L. Lewis asked Donald Henderson, a former college professor and a member of the Communist Party, to organize food growers and processors into an international union. Henderson traveled across the country, recruiting farmers, pickers, canners, and packers as members of the United Cannery, Agricultural, Packing, and Allied Workers of America (UCAPAWA).[29] Kester and Mitchell had serious reservations about the new union but agreed to go along with it because "it seemed to be the logical wave of the future."[30]

Whitfield attended the founding convention of UCAPAWA in the summer of 1937. In an article published in the *St. Louis Post-Dispatch* in August of that year, Snow described an earnest conversation with Whitfield about the affiliation of the STFU with the CIO. Snow "had read that this Denver conference was coming off, and had seen only sketchy press accounts of its proceedings." He wanted to know more about it, and although he realized it was not "exactly cricket for a planter to listen at length to a cropper," he contacted Whitfield.[31]

Snow described Whitfield as a "dusky, bald-headed and enthusiastic cotton-cropper." Another contemporary account confirmed that Whitfield was a lean, "coffee-colored" man with a lined faced

28. Schroeder and Lance, "John L. Handcox," 126.
29. Lichtenstein, introduction to Howard Kester, *Revolt among the Share-croppers*, 43–44.
30. Mitchell, *Mean Things Happening*, 165.
31. Thad Snow, "Why Share-Croppers Join the CIO."

"surmounted by an impressive bald cranium." The story of his bald-
ness, recounted in Snow's article, was that Whitfield, as a young
man, had tried to follow the current fashion of straightening his
hair. One day he purchased a bottle of a hair-straightening agent
from a traveling salesman. The chemical destroyed his hair and
blistered his scalp. No hair ever grew back. Snow offered this story
as a humorous anecdote, explaining why a man in his forties looked
much older than his years.[32]

Whitfield's name never appeared in the 1937 article; the labor
leader remained an anonymous "darky." Snow implied that Whitfield and he had not previously met, when he wrote, "So when I
heard in a roundabout way that a Mississippi County darky had actually attended [the convention], officially representing the Missouri
division of the Southern Tenant Farmers' Union, I forgot my dignity
and sent for him. Surely, if our cotton workers are going to organize
on us, we planters ought to know what is going on."[33] This statement was not strictly factual in light of substantial evidence that
Snow and Whitfield were in contact as early as the spring of 1936.

Addressing a white audience, Snow was apparently not completely comfortable admitting his association with a black man. His
language was patronizing and racist, but his collegial feelings toward Whitfield were genuine. The association lasted throughout
Snow's life.[34] Long after his death, Snow's daughter referred to the
white farmer and the black minister as friends and colleagues.[35]

Why did he not reveal Whitfield's name? This could be construed
as demeaning—an attempt to keep the black man obscure and anonymous. But it probably reflected a desire to protect the labor movement, and the activist himself, from possible acts of retribution.
These were dangerous times, and Whitfield's message was con-
frontational.

Instead of waiting patiently for God to help the poor farmers,
Whitfield told the poor farmers to help themselves. "I was just full

32. Ibid.; Belfrage, "Cotton-Patch Moses," 94.
33. Snow, "Why Share-Croppers Join."
34. Barbara Whitfield Fleming recalled that she and her father attended
Snow's book signing at the Famous-Barr store in St. Louis in 1954 (telephone
interview with the author, September 10, 2002).
35. Delaney interview, August 16, 1999.

of dynamite," he told a reporter, "from head to foot." He announced
to his listeners that "anyone [who] can tell you about Heaven and
can't tell you how to get a loaf of bread here—he's a liar." He believed
that God had blessed this country with enough food to feed all the
hungry people. "Somebody's gettin' it," he said to the croppers. "If
you ain't, that's your fault," not God's.[36] He believed that if farmworkers banded together in a union, they could get what they needed.

Snow understood these sentiments, but he also liked and admired Whitfield in a personal way. Initially, it was the preacher's
sense of humor that beguiled him. Whitfield's daughter Shirley
remembered her father as a fun-loving man. "He was very charismatic," she said. "People just liked him right away."[37] Apparently, Whitfield felt comfortable enough to tell Snow the story of
how he lost his hair, and Snow appreciated a good anecdote. Over
the course of several years, Whitfield stopped at Snow's Corner
many times to tell stories and to talk about the sharecropper system, the AAA, and the problems of making a living in the cotton
country.[38]

As a man in his fifties, Snow was impressed by the younger man's
energy and drive. Whitfield returned from the Denver conference
with "freshly-kindled aspirations," which he confided to Snow.
Clearly, Snow respected and perhaps envied the "organizing zeal"
that powered Whitfield to keeping "going day and night talking at
union meetings."[39] The younger man's energy must have been particularly poignant to Snow, a grieving widower who spent much of
his time in bed reading Veblen and suffering from an unexplained
paralysis.

In 1938, Whitfield's outspoken support of the union led to his
eviction from the farm where he worked. Snow would have accepted him as a tenant, had he asked. But on the day before Whitfield and his family would have become homeless, he learned that a
new, painted, weatherboarded house was waiting for him at La
Forge. According to Cedric Belfrage, a British journalist who reported on events in the Bootheel, Snow used his connection with

36. Belfrage, "Cotton-Patch Moses," 97.
37. Farmer interview, April 7, 2002.
38. Delaney interview, August 16, 1999.
39. *St. Louis Post-Dispatch*, August 9, 1937.

Hans Baasch, who helped run the progressive La Forge communities, to make this possible.[40]

Snow understood the danger of Whitfield's union activities. Other local planters chose not to be concerned. "We have not even troubled to break up and terrorize union meetings," Snow wrote.[41] But he avidly read accounts of the troubles in Arkansas. He believed that a great conflict was coming and that planters would not surrender without a struggle.

Unlike other planters, he considered the situation from the share-croppers' point of view. The STFU arose, he believed, from the "budding industrialization of cotton-growing." By "industrialization," he meant the increasingly businesslike relationship between planters and croppers, "the stiffening by planters of the share-crop terms," and the trend toward increasing use of day labor in the cotton fields. He realized that sharecroppers, however economically deprived they were, had an attachment to the land and a stake in the success of the cotton crop. Day laborers were merely transients with no roots in the soil and little to lose by joining the union. But share-croppers and even rent-paying tenants also joined the union, for they could understand the trend; they could imagine themselves being reduced to the status of day laborers.[42]

In many ways the most significant threat to the sharecroppers was the mechanization of cotton picking. By the summer of 1936, inventors had developed a tractor-powered cotton-harvesting machine. In August of that year, representatives of the USDA, planters, and dealers watched as the picker moved down rows of cotton, stripping the fiber from the bolls onto a set of moistened spindles on a revolving drum. Officials predicted that the picker would reduce picking costs about two-thirds by reducing the need for hand pickers. Howard Kester viewed this machine as more frightening than the planters' reign of terror. Snow wrote of tenants who joined the union: "I suppose they sense the threat of complete industrial-

40. Belfrage, "Cotton-Patch Moses," 99. Shirley Whitfield Farmer remembered seeing Belfrage in meetings with her father and Snow at Cropperville. She also recalled that Belfrage was sent out of the United States for being a communist. Farmer corresponded with him until he died of a stroke about 1990 (interview, April 7, 2002).
    41. *St. Louis Post-Dispatch,* August 9, 1937.
    42. Ibid.

ization, with possibly the mechanical cotton picker in the offing, and feel the need of solidarity."[43]

Snow also understood the importance of racial solidarity in the union. With his point of view as a planter, he could see that racial animosity had divided and weakened the sharecroppers. For white and black members of the STFU, the organization represented "economic unity strong enough to neutralize their long-burning race antagonism."[44] Planters had often clung to their power by playing white workers against black workers when there was any threat of a labor uprising. A strong and vigorous union could undercut this advantage and give sharecroppers a fighting chance at a fair deal.

An alliance between the STFU and the CIO had the potential to alter the balance of power in cotton country. Snow believed the introduction of industrial unionism in the rural South was the logical result of the threat of mechanization in the cotton fields. In 1937, however, he could not know that within two years UCAPAWA and the STFU would split apart in acrimony and confusion, that Whitfield would shift his loyalties to UCAPAWA, and that the sharecroppers of the Bootheel would participate in a massive and unprecedented sit-down strike along two major highways. At that time, Snow knew only that farmworkers were engaged in a monumental struggle and that "no one [could] guess when and how the inevitable changes [would] come on account of it."[45]

The frontier farmer of 1910 had made a long ideological journey to become a CIO sympathizer in 1937. What brought Snow to this point? Apparently there was not one particular event that radicalized him. In the 1920s, he made the transition from the midwestern style of farming, which used hired hands, to the southern-style of cotton plantation, which used sharecroppers. The change in identity from "farmer" to "planter" disturbed him. By the end of the decade, he had faced a personal financial crisis and witnessed the stock market crash. The Great Depression jolted him. Although he escaped personal ruin, he struggled with the problem of dealing fairly with his laborers. He began to listen to what the workers were saying, and he met the charismatic Whitfield.

43. *Cape Girardeau Southeast Missourian*, August 31, 1936; Kester, *Revolt of the Sharecroppers*, 86; *St. Louis Post-Dispatch*, August 9, 1937.

44. *St. Louis Post-Dispatch*, August 9, 1937.

45. Ibid.

At this point in his life, Snow was a troubled man who sought temporary escape from the tensions of the Bootheel. In early 1938, he wrote a letter to his friend Charles Ross, in which he suggested the possibility of social revolution. For much of the ensuing year, he and his two teenaged daughters traveled in Texas and Mexico, where he rested, swam, and slowly recovered from his emotional and physical ailments. His youngest daughter, Emily, who was clearly his favorite, shared his political philosophy and became his confidant, probably helping to fill the void left by her mother's death. He gradually gained strength and balance and was able to walk and drive, although he walked with a limp for the rest of his life. Late in the summer, he gave up his cane and returned home, ready to tend to his farm.[46]

He returned to the Bootheel with renewed energy and a sharpened sense of outrage. On the trip back from Mexico, he stopped in county seats all across the United States and spoke with local officials about federal agricultural policy and how it affected the people who were trying to earn a living from the land. He found that people were willing to talk to him because many of them were as dissatisfied as he was with the way things were going. In the end, he concluded that all across the country, local administrators and landowners were pushing tenants and sharecroppers off the land and denying them the benefits of New Deal programs. This, he believed, "amounted to a national scandal."[47]

At the end of October 1938, he sent a revealing letter to the editor of the *St. Louis Post-Dispatch*. In it he reflected on his experiences in Mexico, where he said he became an admirer of President Lazaro Cardenas. Snow spent much of his time lying on a beach in Acapulco, talking politics with members of Mexico's upper class. Most of these men opposed Cardenas's political program, which included agrarian reform and redistribution of land.[48] But, Snow said,

46. Snow, *From Missouri*, 225–26. According to Fannie Snow Delaney, during this trip, her father visited Leon Trotsky in Mexico. She recalled waiting in the car with her sister Emily while their father went into the house with him. She said Trotsky came out of his house, walked across the yard to the car, leaned in, and talked to the two teenage girls. Then, she said, her father went into the house with him (Delaney interview, August 16, 1999).

47. Snow, *From Missouri*, 228–30.

48. *St. Louis Post-Dispatch*, October 27 and 31, 1938.

there was an important difference between Cardenas's enemies and Roosevelt's. In Snow's view,

> The upper-class Mexican business and professional men hate the Mexican New Deal (they are for Franco in Spain, too) just like our best people hate our New Deal, but they don't hate Cardenas like we hate Roosevelt. Maybe it is partly because Cardenas rose from the peons and can't be hated, as it is said Roosevelt is hated by Liberty Leaguers for his betrayal of his class.[49]

When Snow said "our best people," he meant it ironically to designate the economic elite. By using the pronoun "we," he seemed to associate himself with this group. But previously in the same letter, he made a point of saying that he had talked not only with the wealthy elite, but also with members of Mexico's lower class, "all of whom adored" Cardenas. Because of this, Snow wrote, "I had to give him my affection as well as my respect."[50] In this brief but complex letter, Snow brought to light his own divided loyalties—his identification with and detachment from the moneyed classes, his attraction to and distance from the poor.

During the 1930s, Snow contributed repeatedly to the *Post-Dispatch*'s "Letters from the People" column. By the end of the decade, he had become the paper's "foremost amateur contributor," receiving a large volume of fan mail. He wrote fluently and often humorously on many subjects, including hunting, the economic problems of farmers, Veblen's ideas, interesting local characters, and social conditions in the Bootheel. Although he remained a dedicated farmer, a reporter commented that "Thad is reputed to spend only an hour a day on farming and the rest on reading, writing and studying."[51]

Despite his radicalism, Snow remained a member of the leisure class. He could espouse causes because he did not have to expend all his time and energy scratching out a living. He sympathized with his laborers, but he was separate from them. A contemporary observer described his attitude toward black sharecroppers as "paternalistic." He felt obligated to take care of them. "Sharecroppers

49. Ibid., October 27, 1938.
50. Ibid.
51. *St. Louis Post-Dispatch*, May 2, 1937.

who came to his place remained there to die of old age."[52] He was kind to them, but it was the kindness of a superior to an inferior. Although he gave moral support to the STFU, he could not join the union. He was still a landowner, and he was still caught up in a system that gave landowners' the privilege of exploiting the labor of others. To his credit, he viewed the situation with some level of detachment, realizing that his own personal interests might not emerge triumphant in a contest with a militant working class. Self-mockery and a sly realization that justice might just prevail were evident when he wrote, "Well, if it comes about that I can no longer get my labor for practically nothing, I may at least observe with active interest the progress, mistakes and achievements of a militant farm labor organization."[53]

His public support for the farm laborers was, by any account, unusual behavior for a large landowner. Why did he do it? The simplest explanation would be that he had a compassionate nature. Recall that Snow had picked up Nellie Feezor Stallings and her siblings—the children of farmworkers—from the country school on rainy days to give them a ride home in his wagon. This may have been simply a kindness extended by a patrician to members of the lower class, for Snow had not sent his own children to the country school, but paid tuition for their education in Charleston. Another possible explanation would be that he could sympathize with hard-pressed farm laborers, having suffered his own personal tragedies.

Intellectually, he was a man who looked for answers from a wide array of sources, including Baptist ministers, Bible salesmen, union organizers, and social philosophers. By the spring of 1937, before Lila's death, he had become a devotee of Veblen,[54] whose *The Theory of the Leisure Class* condemned the American economic elite as selfish and boorish, conspicuously enjoying comfort and opulence at the expense of the workers. It was clear that Snow embraced this message when he blamed the troubles of the sharecroppers, at least in part, on the stubborn self-interest of the planter class.

The fact remains that he continued to benefit from the labor of the croppers on his farm. His support for the STFU remained only a

52. Belfrage, "Cotton-Patch Moses," 98.
53. *St. Louis Post-Dispatch,* August 9, 1937.
54. *St. Louis Post-Dispatch,* May 2, 1937.

gesture—an unusual one for a planter—but only a gesture. In order to win justice, farmworkers would have to take matters into their own hands.

While Snow dealt with his personal troubles and his intellectual travail, the sharecroppers continued to suffer. Whitfield settled his family in a snug house at La Forge, but he did not stop fighting for justice. His wife, Zella, supported his decision to continue speaking at union meetings and planning a large-scale protest.[55] He knew that projects like the one at La Forge helped only a small number of those in need. Without solidarity, the vast majority of farm laborers would not make a decent living or receive a fair shake from the landowners.

One kind-hearted and intellectually conflicted planter could not bring justice to the workers. Handcox's poem "Out on Mr. Snow's Farm, or The Kind of Man We Like to Meet," expressed appreciation of Snow's support and sympathy. But other songs Handcox recorded for the Archive of Folksong in 1937, such as "Raggedy, Raggedy Are We," expressed a deep sense of rage at the sharecroppers' suffering:

> Hungry, hungry are we . . .
> Landless, landless are we . . .
> Homeless, homeless are we . . .
>
> Cowless, cowless are we,
> Just as cowless as cowless can be
> The planters don't allow us to raise them,
> So cowless, cowless are we . . .
>
> Pitiful, pitiful are we
> Just as pitiful as we can be
> We don't get nothing for our labor
> So pitiful, pitiful are we.[56]

This song blamed the sharecroppers' plight directly on the planters. It was not the system that made them suffer; it was the planters'

55. Belfrage, "Cotton-Patch Moses," 99. According to Shirley Whitfield Farmer, her mother "worked with Daddy a lot, but she was mainly home with the kids" (interview, April 7, 2002).
56. Schroeder and Lance, "John L. Handcox," 127–28.

greed, pure and simple. The planters did not allow the croppers to raise their own livestock, because that might give them too much independence. The planters paid them little or nothing for their labor, and because of this they were pitiful and hungry. The planters evicted them, and so they were homeless.

Another famous song by Handcox, "We Gonna Roll the Union On," expressed the determination of farmworkers to carry on with or without support from friendly planters and capitalists:

> We gonna roll, we gonna roll, we gonna roll the union on.
> We gonna roll, we gonna roll, we gonna roll the union on.
> If the planter's in the way, we gonna roll it over him,
> Gonna roll it over him, gonna roll it over him.
> If the planter's in the way, we gonna roll it over him.
> Gonna roll the union on.[57]

---

57. Ibid., 129.

The Great Roadside Demonstration 6

Mr. Snow told me that he was in sympathy with the sharecroppers and that he was studying Social conditions in the U.S. and was interested in the problems of these people. He stated that this was a [good] demonstration and some one had done a good job of organizing these people.

—Lt. Col. Harry E. Dudley, 140th Infantry Regiment
Missouri National Guard, to Governor Lloyd C. Stark

According to Snow, the Sharecroppers' Roadside Strike of 1939 originated with a forlorn joke by a hard-pressed man. Sharecroppers often faced eviction in the winter months when their labor was no longer necessary. At a meeting of one of the locals of the STFU, one cropper reported that he would have to leave his cabin. Asked where he would go, he replied that he would move to the roadside. In Snow's retelling of the story, Owen Whitfield conceived from this remark the idea of a mass migration to the state highways as a protest against evictions of tenant farmers.[1]

In the fall of 1938, Snow returned to Mississippi County from his long sojourn in Mexico and the Southwest. He resumed his association with Caverno and other landowners. In September, Governor Stark appointed Snow and Caverno as delegates from Missouri to meet with senators from the cotton-producing states in Washington, D.C., for the purpose of discussing ways to obtain increased subsidies from the federal government. Snow also resumed his

1. *Sikeston (Mo.) Standard*, February 24, 1939.

association with Whitfield, whom he later described as "bursting with enthusiasm."[2]

Early in January 1939, Whitfield spoke with Congressman Orville Zimmerman at his home in Kennett, Missouri, ninety miles south of Charleston, about plans for a sit-down strike on the roadsides of the Bootheel. Zimmerman, who refused to believe the sharecroppers would camp out on the highways in the middle of winter, advised Whitfield to speak to Snow.[3] Apparently, Snow and Zimmerman immediately made a trip to Jefferson City and informed the governor about the planned demonstration.[4]

On the following Saturday night, January 7, Snow and Sam Armstrong, a reporter for the *St. Louis Post-Dispatch,* attended a big meeting at a church in the Sunset Addition, the African American neighborhood on the west side of Sikeston. Whitfield addressed a crowd of about 350 sharecroppers, who laughed at his humorous fables about croppers, planters, and white folks.[5] Snow was surprised at the tone of the sermon, which was not a solemn exhortation, but a rollicking narrative, incorporating "much symbolism, drawn from the Bible and from the habits of wild birds and animals, always with a humorous turn."[6] Snow and Armstrong laughed and cheered with the rest of the congregation, losing themselves in the sense of participating in a great adventure.

Writing in *Harper's Magazine,* Cedric Belfrage later paraphrased Whitfield's sermon. In a sonorous voice, with no need of a microphone, Whitfield invoked the image of the Israelites escaping from slavery in Egypt:

And Moses . . . got 'em to the Red Sea and they made camp there. But here came old boss Pharaoh's ridin' bosses in their chariots. And Moses raised his hand, and the waters parted, and the children of Israel walked across on dry land. . . . We're gonna make an exodus likewise! It's history repeatin' itself in 1939.[7]

2. Governor Lloyd C. Stark to Hon. Olin D. Johnston, Governor of South Carolina, September 23, 1938, Stark Papers, folder 1872; Snow, *From Missouri,* 246.
3. Snow, *From Missouri,* 247.
4. Harry E. Dudley to Hon. Lloyd C. Stark, January 20, 1939, Stark Papers, folder 1959.
5. *Sikeston (Mo.) Standard,* January 10, 1939.
6. Snow, *From Missouri,* 249.
7. Belfrage, "Cotton-Patch Moses," 94.

When he asked the assembled crowd, "How many of you got notice to move," hands went up throughout the church, and people responded, "Me, me!"

"How many got a place to live?"

Hands went down.

"Well," said Whitfield, "You take a turkey or a goose—anything that flies—has to squat first. Where we goin' to go?"

According to Belfrage, half a dozen people responded, "Sixty-one highway!" And the cry was repeated from pew to pew.[8]

Snow regretted later that no one had made a recording of the speech, which he compared to the Sermon on the Mount. Armstrong, the reporter, said it was the best oratory he had ever heard and described it on the front page of the next day's paper.[9]

On January 10, the *Sikeston Standard* reported that hundreds of families who had been ordered to leave their plantations after January 1 might stage a mass demonstration on U.S. Highway 61. The paper quoted Sam Armstrong's *Post-Dispatch* article in predicting that as many as seventeen hundred people might be involved. On the morning of January 11, Zimmerman sent a telegram to Governor Stark, advising him, "Four hundred evicted sharecropper families are camping on highways in Southeast Missouri with no place to go and without food or shelter. Situation is critical and reputation of state is at stake." That same morning, J. R. Butler, president of the STFU, asked Governor Stark to provide tents for the people on the roadsides. Stark replied that the state owned no tents and therefore could not comply with the request.[10]

Throughout the day on January 11, more families turned up on the roadsides. Neat piles of sawn and split stove wood appeared in each of the camps. Reporters from Chicago, Baltimore, New York, and points in between wandered the highways, talking with the croppers, to the exasperation of many local citizens. According to Snow, "Mass hysteria was getting under way."[11]

Colonel B. M. Casteel, the superintendent of the State Highway

8. Ibid.

9. Snow, *From Missouri*, 250.

10. *Sikeston (Mo.) Standard*, January 10, 1939; Orville Zimmerman to Hon. Lloyd C. Stark, January 11, 1939, Stark Papers, folder 1958; Lloyd C. Stark to J. R. Butler, January 11, 1939, Stark Papers, folder 1958.

11. Snow, *From Missouri*, 242, 256.

Patrol, went to the Bootheel and reported back to the governor on January 12 that there were about 250 families, comprising about twelve hundred individuals, in thirteen camps along Highway 61 from Sikeston to Hayti and on Highway 60 from Sikeston to Cairo. The colonel had eight men assigned to twenty-four-hour duty, keeping the demonstrators out of the main roadways. Casteel asserted that most of the people were migrant workers from neighboring states. He believed that the leaders responsible for bringing them into Missouri were "Thad Snow, large land owner of Charleston, Missouri; [Baasch] of the La Forge Project; Whitfield, negro preacher."[12]

The Sikeston newspaper described the people who were suddenly appearing on the roadsides. One camp near the Y-bridge across the Mississippi River on Highway 60 held about a hundred people from the Dorena area south of Charleston. Two-thirds of the people were black. According to the newspaper, "Families ranged from five and six persons up to a couple with thirteen children. All were garbed in the rough clothing of the sharecropper." One bystander asked if the babies and children had milk to drink. Another onlooker responded that "the only milk given the babies was titty milk and when that gave out they were put on black coffee."[13]

Four miles west of Charleston, at the junction of Routes 60 and 55, there was a much larger camp, where about 250 people had taken up residence. This was a busy intersection, and motorists sometimes stopped to ask what was going on. A spokesman for the group replied briefly and politely that the boss man had ordered the people to leave the plantations on January 1. The law gave them ten days' grace. After January 10, they were homeless. They had been informed that they had the right to be on the roadsides as long as they did not bother passing cars.[14]

Most of the demonstrators were black sharecroppers. The majority of the white families who chose to be part of the demonstration congregated in a large camp just south of Sikeston on Route 61. According to Snow, the sight of white families joining in a black protest shocked the people of Sikeston more than any other aspect

12. Les Forman, Memorandum, to Governor Stark, January 12, 1939, Stark Papers, folder 1958.
13. *Sikeston (Mo.) Standard,* January 13, 1939.
14. Snow, *From Missouri,* 242.

of the protest. In *From Missouri,* he wrote that "It was incomprehensible and therefore sinister to every right-thinking observer and defender of the social order."[15]

Cleve Mattox, a white sharecropper, told a reporter for the *Poplar Bluff Daily American Republic* why he and his family joined the demonstration. He was forty-six years old, he said, and had "been farming ever since he could lift a hoe." He received a sheriff's notice to vacate the farm where he worked:

> The man me and my son worked for was a renter and we were his only sharecroppers. He told us he didn't want any more sharecroppers, he was going to handle it all himself, with day labor.
>
> Lots of folks are doing that because they just pay day labor six-bits [seventy-five cents] a day whenever they need it, and they don't have to share the crop or the government benefits with anyone.[16]

Because he had nowhere else to go, he brought his family out to the highway.

How did all those demonstrators get out to the roadsides during the night of January 9–10 without attracting attention? According to the *Sikeston Standard,* black and white families came in old trucks and cars, some with a cotton sack full of supplies, others with furnishings for a small house.[17] However, no cars, trucks, or wagons, were parked beside the camps. Most of the sharecroppers had no means of transportation. Other people—friends and supporters—must have brought them there.

When the demonstrators first appeared on the roadsides, they had no fuel, no food, and no shelter except for quilts and blankets draped on poles.[18] Each night for nearly a week, a caravan of vehicles appeared and vanished, leaving supplies at the campsites. Croppers who had not been evicted brought provisions to their friends and relatives. For every family in public view, there was someone behind the scenes, supporting them physically and in spirit.[19]

---

15. Snow, *From Missouri,* 243; see also Strickland, "Plight of the People," 410.
16. *Poplar Bluff (Mo.) Daily American Republic,* January 17, 1939.
17. *Sikeston (Mo.) Standard,* January 13, 1939.
18. *Sikeston (Mo.) Standard,* January 20, 1939.
19. Snow, *From Missouri,* 243–44.

Snow visited the camps often. In an excited mood, he traveled around the countryside, talking with groups of friends and with many of the campers. He felt healthier than he had in years, and his legs seemed stronger. When he talked with campers, they told him they were not worried about running out of food and fuel. They knew that he knew that some of the croppers on his farm used old cars and trucks to collect and deliver provisions at night. They cut trees in his woods for fuel and took gas for their vehicles from his tank. Sometimes when they came in the night to fill up, he would turn on the light for them. Odie Reeves, a farmer in southern Mississippi County, reported to state officials that Snow sent quilts and blankets to the camps on the highway.[20]

Other white middle-class people quietly offered aid, too. According to Fannie Cook, who came from St. Louis to study the situation, many white citizens, including doctors, housewives, and café owners, sympathized with the croppers. In a letter to her friend Edna Gellhorn, she wrote that "Thad Snow is by no means the only planter who feels as he does, but the others are afraid to speak out."[21]

The Sikeston newspaper quickly pointed out that Whitfield was not present in the roadside camps. Colonel Casteel also reported that the minister, presumed to be the leader of the demonstration, was "missing."[22] Describing him as an "agitator," newspaper editor Charles L. Blanton urged that he be evicted from his house at La Forge. "Besides," Blanton commented, "He was not entitled to one of these farms as he was formerly one of Thad Snow's tenants over in Mississippi County."[23] This assertion was untrue, but it served the purpose of publicly linking Whitfield and Snow. Alarmed by the situation and searching for someone to blame, many local residents concluded that Snow, Whitfield, the STFU, the CIO, or a combination of them all had instigated and planned the demonstrations.

One white sharecropper, quoted in the Sikeston newspaper, alleged that the CIO and the STFU had organized the strike in order

20. Ibid.; H. G. ("Chilli") Simpson to Hon. Lloyd C. Stark, January 21, 1939, Stark Papers, folder 1959.

21. Fannie Cook to Edna Gellhorn, April 17, 1939, Fannie Cook Papers, box 5, folder 2, Missouri Historical Society, St. Louis.

22. B. M. Casteel to Hon. Lloyd C. Stark, January 20, 1939, Stark Papers, folder 1959.

23. *Sikeston (Mo.) Standard*, January 13, 1939.

to attract the attention of the press and force Congress to take action on behalf of oppressed farmers. This came close to the truth, except that unions played only a peripheral role in the roadside strike. By the beginning of 1939, the relationship between UCAPAWA and the STFU had become strained to the breaking point. Whitfield believed the future of the tenant farmers' movement lay with UCAPAWA and the CIO.[24]

STFU leader J. R. Butler apparently depended on the press for information about the planned demonstrations. On January 9, he wrote to the union's cofounder, H. L. Mitchell, saying that the "*St. Louis Post-Dispatch* reports 1700 evicted tenants [and] sharecroppers of Southeast Missouri going on Highway 61 tomorrow. . . . I go tomorrow. . . . Will make demands on Red Cross, WPA, and Governor." The next day, Whitfield sent a curt telegram to Mitchell, saying, "Evicted sharecroppers now moving to highway. You and Butler keep out." The mild-mannered Butler did come to the Bootheel to try and assess the situation. He told local citizens and reporters that he knew nothing of the plans for the strike but was in the region to ascertain what it was all about.[25]

State officials, including Dr. Harry F. Parker, the state health commissioner, arrived to investigate the situation. Beginning on January 12, Parker and Casteel visited various camps, finding unsanitary conditions. In Charleston, on the morning of January 13, they conferred with H. G. "Chilli" Simpson of the State Highway Commission, who assisted them in their investigation. The next day, the Highway Patrol began moving people from the roadsides.[26]

Snow met with Simpson in his office on the afternoon of January 13. According to Simpson, Snow had been traveling around town, expressing "his opinion that this was a fine demonstration to Joe Moore, Art Wallhausen, newspaper editor here, and several others." State troopers reported that Snow had been visiting the camps. In the meeting at his office, Simpson tried to convince Snow that the

24. Donald H. Grubbs, *Cry from the Cotton: The Southern Tenant Farmers' Union and the New Deal*, 180; Strickland, "Plight of the People," 405.

25. J. R. Butler to H. L. Mitchell, January 9, 1939, reel 10, Southern Tenant Farmers' Union Papers, 1934–1970 [microfilm]; Whitfield to H. L. Mitchell, January 10, 1939, STFU Papers, reel 10; *Sikeston (Mo.) Standard*, January 13, 1939.

26. B. M. Casteel to Hon. Lloyd C. Stark, January 20, 1939, Stark Papers, folder 1959.

majority of people in Mississippi County believed that the demonstrators should be removed from the highways and returned to the
farms from which they came. During the meeting, Colonel Harry E.
Dudley of the Missouri National Guard came to Simpson's office.[27]

Snow asked Simpson to confer with T. J. North, a black man, who
was one of the leaders of the demonstrations. Snow and Dudley
searched for North at the roadside camps. Failing to find him, they
returned to Charleston. According to Dudley, "Stopping at a Negro
Café, owned by Marshal Kern, Mr. Snow met T. J. North and Kern
and suggested to them that they meet with Mr. Simpson at his office. They agreed to do so. Kern stated that 25 negro women and
children stayed in his café during the night, returning to the
Highway during the day."[28]

In the presence of Simpson, Dudley, North, and Kern, Snow
talked about a woman named Annie, who had left his farm to join
the demonstrations. Dudley and Simpson gave divergent accounts
of Annie's story. According to Dudley, Snow said that he "had a
negro woman named Annie who did not have to move off from his
farm but that she was in sympathy with the movement and that
she had a perfect right to stay on the roadside and keep up her
demonstration. He added that Annie was a little thick headed you
know." Simpson reported it this way: "Mr. Snow stated that he had
a negro woman on his farm, by the name of Annie, who had a perfect right to move out if she desired to do so, although she had not
yet done so."[29]

In his letter to Governor Stark, Simpson offered another interpre-
tation of Snow's story about Annie. According to Simpson, Annie
and a black man named Jake Reed left Snow's farm, not only out of
sympathy with the strikers, but also because of a grievance against
Snow. Reed and Annie apparently told Melvin Dace of the State
Highway Patrol that they left the farm because Snow had demanded that they sign over their parity checks to him.[30] Simpson
offered no corroboration for these allegations. While there is no de-

27. H. G. ("Chilli") Simpson to Gov. Lloyd C. Stark, January 21, 1939, Stark
Papers, folder 1959.
28. Harry E. Dudley to Hon. Lloyd C. Stark, January 20, 1939, Stark Papers,
folder 1959.
29. Ibid.; Simpson to Stark, January 21, 1939.
30. Simpson to Stark, January 21, 1939.

finitive proof either way, such an action by Snow seems highly in-
consistent with letters he wrote between 1934 and 1938, fretting
about the possibility that he had inadvertently misled his sharecroppers and criticizing unfair treatment of farmworkers.[31]

This was a terrible allegation against a man who positioned himself as a friend of the roadside demonstrators. Clearly, Simpson was
perturbed with Snow. In the first paragraph of his letter to the governor, Simpson stated that "The people of this section were up in
arms, not only landowners, but merchants and people in every walk
of life." He went on to report that he had seen Snow on the street the
day before the demonstration, and Snow had known about it then
and approved of it. When Snow brought North and Kern to his office, Simpson tried to persuade them to bring the demonstrations to
an end. However, he reported to the governor, "At this time Mr.
Snow spoke up with 'Let them set out there if they want to,' and I
could see that he immediately threw a damper on the plans. The
meeting broke up, Snow still protesting that they had a right to go
out on the road and make their demonstration." Simpson viewed
Snow as an obstructionist. Dudley reported, however, that Snow
agreed that the demonstration had accomplished its goals and that
the people should return home.[32]

Simpson gave credence to allegations against Snow, although he
discounted similar accusations against other landowners. In general, he doubted that any of the demonstrators had actually been
evicted. Of more than one hundred families questioned, he doubted
"that there was one legal eviction notice." In particular, he discounted
statements from black sharecroppers, writing that "Several of the
negroes, when questioned, stated that they had been moved from
farms, and, upon investigation, their statements were found to be
incorrect."[33] While Snow was not necessarily innocent, according to
Simpson, he was effectively the only Bootheel landlord who had
mistreated his sharecroppers.

On the morning after the meeting in Simpson's office, officials
began removing the demonstrators from the roadsides, using trucks
provided by the State Highway Department. One week after the

---

31. See in particular Snow to Anderson, October 20, 1934.
32. Simpson to Stark, January 21, 1939; Dudley to Stark, January 20, 1939.
33. Simpson to Stark, January 21, 1939.

roadside strike began, the *Sikeston Standard* reported that state and county officials had moved thirteen hundred people between Saturday, January 14, and Monday, January 16.[34] Some of the sharecroppers returned to their old farms, but many landowners refused to take them back.[35] In New Madrid County, numerous families could not return to their former homes. Officials transported them to a camp on a forty-acre tract of public land behind the levee about six miles east of La Forge.[36]

Dispersed from the highways, the sharecroppers found refuge in various temporary encampments. The swampy land behind the levee in New Madrid County became known as "Homeless Junction." Some of the dispossessed farmers took shelter in a rundown dance hall in Charleston. Walter Johnson became the leader of a group of refugees in a community centered around the Sweet Home Baptist Church near Wyatt. Some of the people there slept on pews. Others camped in the yard. William R. Fischer took charge of a white camp near Dorena, where food was in short supply. Other families crowded into tents and shacks near Hayti and Charleston.[37]

Lorenzo J. Greene, a history professor from Lincoln University in Jefferson City, came to the Bootheel to address faculty, students, and citizens at Charleston High School after the demonstrators had left the highways. In the days following his lecture, he visited several of the temporary camps, including the Sweet Home Church camp, and was shocked by what he observed. As he described the situation:

> Hundreds of sharecroppers with their pitiable belongings congregated in groups. Men, women, young and old, boys and girls shivering in the cold, little children and even babies crying because of hunger, their swollen bellies indicative of lack of sufficient food. I saw their makeshift dwellings of wood, burlap, tin, cardboard, anything to protect them from the frigid weather. I saw them trying to cook over open fires, on makeshift stoves, or just standing about trying to keep warm.[38]

34. *Sikeston (Mo.) Standard,* January 17, 1939.
35. Dudley to Stark, January 20, 1939.
36. *Sikeston (Mo.) Standard,* January 17, 1939.
37. Mitchell, "Homeless," 9; Belfrage, "Cotton-Patch Moses," 101; Greene, "Lincoln University's Involvement," 27; *St. Louis American,* February 16, 1939.
38. Greene, "Lincoln University's Involvement," 27.

Upon his return to the classroom, he inspired students from Lincoln to collect money, food, and clothing for the sharecroppers.

Strike leaders set up headquarters in St. Louis, where some were reportedly "hiding out" after receiving death threats. One of them, W. P. Wells, told a reporter for the *St. Louis American* that the state police carried him back to his old farm near Matthews when they cleared the highways. Planters there seized his livestock and refused to pay him for his crops. He also said that he and his family received death threats. Whitfield found a safe haven at 420 New York Avenue in Kirkwood, where the *St. Louis County Directory* listed him as a resident from 1939 through 1943.[39]

From his suburban home base, Whitfield traveled around the country, trying to win allies for his cause. He went first to St. Louis, and then, during the last two weeks of January and the first two weeks of February, he met with officials and groups in Chicago, New York, Washington, D.C., and Memphis. During this time, Zella and their children, ranging in age from one month to twenty-four years, remained in the home at La Forge.[40]

In St. Louis, Whitfield enlisted the enthusiastic support of Fannie Cook, who had recently served as chairman of the Race Relations Committee of the city's Community Council. When news of the demonstrations appeared in the national papers, friends called and asked her to take some action. She visited the offices of the St. Louis Urban League and there, by chance, met Whitfield.[41]

Cook and the Pulitzer Prize–winning novelist Josephine Johnson worked together to help form a coalition of support groups, including the Urban League, Eden Theological Seminary, Fellowship of Reconciliation, and UCAPAWA. Representatives of these organizations joined to create the St. Louis Committee for the Rehabilitation of the Sharecroppers, which later incorporated as the Missouri Committee for the Rehabilitation of the Sharecroppers. On January 22, the group sponsored a mass meeting, attended by six hundred people, at the Amalgamated Clothing Workers Union Hall in St. Louis to raise awareness of the sharecroppers' plight and also to raise money for relief.[42]

39. *Polk's St. Louis County Directory,* 1939, 1941, 1943.
40. *Sikeston (Mo.) Standard,* February 24, 1939.
41. Cadle, "Cropperville," 28–30.
42. Ibid., 30.

Planters in the Bootheel demanded a federal investigation into the demonstrations. Twenty-seven landowners and landlords from New Madrid, Dunklin, Pemiscot, and Scott Counties signed a petition on January 12, charging Hans Baasch and Owen Whitfield with clandestine and subversive activities. Specifically, the petition charged that

> Whitfield has been going about Southeast Missouri collecting dollars from the poor people of these communities and telling them that if they would move out on the highways the Government would give them forty acres of ground and the tools to cultivate it and that the said Hans H. Baasch is reliably reported to have made various communistic remarks leading to this trouble.[43]

The landlords' petition did not name Snow, perhaps because, despite his opinions, he was a landlord himself. Copies of the petition went to Governor Stark, Senator Harry S. Truman, Senator Bennett Champ Clark, and other officials. Landlords in Charleston held meetings and expressed similar feelings about the demonstrations, which they believed were well planned and carried out by outside agitators and disgruntled day laborers, who had not really been evicted from the land.[44]

In response to the landlords' petition, the FBI did launch a brief investigation. By March 8, 1939, the bureau had concluded its investigation and determined that the strike was a spontaneous expression of indignation by outraged sharecroppers.[45] The STFU was not the sponsor of the demonstration. Landlords may have been surprised by the uprising, but they had mostly their own apathy to blame. The FBI concluded that the landlords underestimated the disaffection of the sharecroppers and misunderstood the situation.[46]

Local rage against Whitfield and his supporters (including Snow) reached a dangerous pitch, fueled by Blanton's editorials in the *Sikeston Standard*. From the beginning, the editor linked Whitfield and

43. Resolution adopted January 12, 1939, at the Court House in New Madrid, Missouri, Stark Papers, folder 1958.

44. *Sikeston (Mo.) Standard,* January 17, 1939.

45. J. Edgar Hoover to Hon. Lloyd C. Stark, March 8, 1939, Stark Papers, folder 1961.

46. *Sikeston (Mo.) Standard,* March 14, 1939.

Snow, identifying them both as left-wing agitators, as in the following comment: "From the best information at hand the organized tenant farmers parades to the roadside of Southeast Missouri has simmered down to the doors of Thad Snow and his former riding boss, the negro Whitfield, both of whom lean to the left very pronounced."[47]

Snow recalled this item as "only mildly suggestive" and not as slanderous or vulgar as most of Blanton's pronouncements. But on the following morning, two carloads of reporters showed up at Snow's door. They told him that gossip was spreading like wildfire, and people were saying that he had planned and managed the roadside demonstrations. He made no statement and refused to deny any allegations.[48] The incendiary editorials continued.

Blanton called himself the "Polecat," because one of his critics had remarked that his writings smelled like a skunk.[49] Enamored of all things southern, he regarded the roadside strike as a threat to the established social order and used his column, with a skunk as masthead, to castigate anyone who did not agree with him. On January 17, he editorialized that Whitfield had duped the "poor negro tenants" and then, in a cowardly manner, "left the country for St. Louis to escape the wrath of the landowners that he had put in such a bad light throughout the United States."[50]

A few days later, the anger burned hotter, and Blanton threatened:

> Things down the line have quieted down in so far as the croppers are concerned as they have vamoused [*sic*] from the highways. Now what will happen, if and when, Whitfield returns to La Forge is another matter. We are afraid the former Ku Kluxers of Southeast Missouri have lost their robes but maybe a few bed sheets or pillow covers would answer the same purpose if they could be revived and parade through and about La Forge.[51]

Within a few days, under the shadow of this threat, the paper reported that Whitfield was gone, but that his family remained at La

47. *Sikeston (Mo.) Standard*, January 13, 1939.
48. Snow, *From Missouri*, 258–59.
49. Dominic J. Capeci Jr., *The Lynching of Cleo Wright*, 140.
50. *Sikeston (Mo.) Standard*, January 17, 1939.
51. Ibid., January 20, 1939.

Forge. One month later, Whitfield's family left the Bootheel and took up residence in Kirkwood, a St. Louis suburb.[52]

Whitfield tried to protect his family from the controversy that swirled around him. His daughter Shirley Whitfield Farmer said there were a lot of things he did not tell his wife and children, so that, if they were questioned, they could truly say they knew nothing of his activities. But he could not insulate them completely. Many years later, Barbara Whitfield Fleming remembered the danger. On her way to school one day, she saw a poster offering a five-hundred-dollar reward to anyone who killed her father. His picture was on it. She also recalled waking up one night when men came to the house in La Forge. She said, "I heard Mama saying, 'He's not here. He's not here. You can search every room but that one. That's my girls' room.'" They searched the house and the barn and then they went away.[53]

Because of Whitfield's association with La Forge, Hans Baasch also came under fire. In its January 13 issue, the *Sikeston Standard* accused Baasch of leaning to the left and hinted that he might have had something to do with the roadside spectacle. On that same day, Snow became worried about Baasch and drove over to Sikeston to see him. His wife and child were frightened, and Baasch himself feared that in the heat of anger someone might take a shot at him on the road to work or at La Forge.[54] Baasch was a sociologist with a foreign-sounding name, and he was associated with the most liberal wing of the New Deal, which did not make him popular with landowners. But he was a paid public official. He remained neutral during the demonstrations, and Whitfield left La Forge when the strike began. After the initial uproar, Baasch was able to go on peacefully with his work.

As gossip spread, the Polecat's editorials became more menacing toward Snow. On January 17, the editor had warned that if agitators continued pulling "stunts" in Southeast Missouri, the KKK would be riding again. Three days later, he accused Snow and Armstrong of distorting facts and stirring up trouble among the sharecroppers. In the next issue, on January 24, he insinuated that Snow was a dangerous lunatic and warned: "We have believed for a long time that

52. Ibid., January 24, 1939; February 24, 1939.
53. Farmer interview, April 7, 2002; Fleming interview, September 10, 2002. See also *St. Louis Post-Dispatch*, February 11, 2001.
54. *Sikeston (Mo.) Standard*, January 13, 1939; Snow, *From Missouri*, 256–57.

Thad Snow was a dreamer, with rattlings in his head and his behind in a twist but the way he has acted of late shows that his head and behind should change places."[55]

In response to these personal attacks, Snow refused to deny his association with the strike or to distance himself from Whitfield. Why did he act so bravely—or foolishly? He knew that the mood in his region was ugly. He knew that Whitfield and perhaps Baasch were at risk, and so he may have tried in this way to deflect anger from them. Years later, he told a friend that he had taken credit, or blame, for the demonstrations in order to absorb some of the wrath against Whitfield.[56] In the initial flurry of alarm and anger, Whitfield's life was certainly in danger.

Snow loved drama and enjoyed being the center of attention. His friend Millie Wallhausen recalled many years later that he also "loved being elusive and mysterious." He reflected in *From Missouri* that he acted on impulse. He was appalled by the panic and fear that transformed reasonable, kind people into "witch burners, torturers, crucifiers, and the like." As he later analyzed his action, "I wanted somehow to show my contempt, not for the people I knew and loved but for the hysterical madness that now consumed them. Perhaps an unworthy trace of exhibitionism still persisted within me, and impelled me to take on the function of Devil when the opportunity appeared. I cannot say."[57]

Some of his former colleagues cast him aside during this crisis. On January 23, Joe L. Matthews, president of the Bank of Sikeston, wrote a letter to Governor Stark, severely criticizing Snow:

> I cannot understand the attitude of Thad Snow in this matter. He has set himself in with the Post-Dispatch and for years has written some articles that sounded fairly good, thus causing a lot of people to have confidence in his ability. I'm frank to admit that I considered him able to get things done along agricultural lines in Washington, but from now on we are all absolutely through with him.[58]

55. *Sikeston (Mo.) Standard*, January 17, 20, 24, 1939.
56. Letter to the author from James L. Lowe, Mountain Home, Arkansas, January 13, 1998.
57. Millie Wallhausen, interview with the author, Charleston, Mo., October 28, 2002; Snow, *From Missouri*, 261.
58. Joe L. Matthews to Governor Lloyd C. Stark, January 23, 1939, Stark Papers, folder 1959.

The Polecat gloated on January 24 that Snow had lost another friend as a result of his actions. Judge Caverno apparently disagreed with Snow's position:

> A beautiful friendship has been torn asunder when Thad Snow went to the left and Xenophon Caverno stayed on the right. Here-to-fore these two gentlemen have always been to the forefront in agricultural matters and were implicitly trusted by landowners and tenants alike, until Mr. Caverno would not go hand in hand with his erstwhile farmer friend.[59]

Despite the newspaper's assertion, Snow's daughter recalled years later that the two men had often disagreed but had remained friends over the years.[60]

Many of Snow's neighbors were incensed at his behavior. The *Sikeston Standard* loved to point this out:

> We were in Charleston for a while Wednesday afternoon seeking atmosphere, as it were, but about all we could absorb was anti-Snow. A number of prominent people who we contacted had nothing good to say about Thad Snow, but all had a dig for him, repeating many stories that wouldn't look good in print. All blamed he and the negro Whitfield for the unenviable notoriety given Southeast Missouri and some in Charleston believed a good horsewhipping would do both of them some good and would be very satisfying to former friends. Anyway, Snow is all melted-up and is a dead cock in the pit.[61]

Snow did not help matters with his facetious "True Confession," published in the *Charleston Courier* on February 3. Neighbors had accused him of masterminding the roadside strikes, and he did not deny this. In fact he confessed to everything and described an out-landish plot concocted in Mexico during an alleged (and perhaps fanciful) meeting with the exiled Russian revolutionary Leon Trotsky.[62]

The "True Confession" said much more about Snow's perverse

---

59. *Sikeston (Mo.) Standard,* January 24, 1939.
60. Delaney interview, August 16, 1999.
61. *Sikeston (Mo.) Standard,* January 27, 1939.
62. Thad Snow, "True Confession???" Snow Papers, folder 32.

nature than it did about the roadside demonstrations. It was a satirical piece of writing that began with a meeting between Snow and Trotsky and progressed through a more and more elaborate and impossible conspiracy story, involving the president of Mexico, President Franklin Roosevelt, Secretary of State Cordell Hull, Upton Sinclair, Al Smith, Norman Thomas, and Snow's little daughter Emily. He offered to show Trotsky's original plan (damaged by saltwater while Emily concealed it in her bathing suit) to "any planter or other local citizen who is interested, and who is sympathetic with the aspirations of the roadside demonstrators."[63]

Some people in the Bootheel may have believed the "confession." But the editor of the *Sikeston Standard* realized that it was a hoax, which proved "conclusively that he would be sent to some asylum." The paper chastised him, with some justification, for trying to "make light of one [of] the most serious affairs that has ever been cooked up in this section."[64] But Snow's misplaced humor was certainly superior to the newspaper's ugly threats against Whitfield and his family.

Despite his "Confession," Snow never doubted the importance of the demonstrations or of Whitfield's leadership. In late February, a group of ministers in Cape Girardeau invited Snow to address an interchurch men's organization on the subject of the sit-down strike. Several hundred men gathered to listen.[65] At that meeting, Snow told the story of how he believed the idea for the demonstration originated—with the sharecropper joking that when he had to leave his farm he would just camp out on the roadside, and Whitfield conceiving from this the plan for a strike that had worldwide repercussions.[66] Clearly, in that meeting, Snow placed the credit for the roadside demonstrations where it belonged, with Whitfield.

After the furor died down, the Bootheel returned to some semblance of normalcy, with its abnormal share of misery for the farmworkers. By the end of January, the sheriff ordered the families camped at Homeless Junction to move back to their old plantations in New Madrid County.[67] But many of the farmworkers had no

63. Ibid.
64. *Sikeston (Mo.) Standard,* February 17, 1939.
65. Snow, *From Missouri,* 271.
66. *Sikeston (Mo.) Standard,* February 24, 1939.
67. *Sikeston (Mo.) Standard,* January 24, 1939.

place to go. Mitchell and other STFU officers proposed the construction of housing for the dispossessed farmworkers on area plantations. In February 1939, the STFU and a delegation of Missouri croppers met with Dr. Will W. Alexander of the FSA and discussed proposals for federally funded housing projects in the Bootheel.[68]

In mid-February, high water flooded out the log cabin of Orange Riggly, who leased land from Snow in Mississippi County. The *Sikeston Standard* published a photograph of Riggly and his family, standing on the wooden porch of a house that looked like an ark in a deluge. Clearly, the newspaper wanted to make the point that Snow, who had become the white champion of the sharecroppers, ought to tend his own affairs.[69]

One month later, Fannie Cook asked him to introduce her to the world of sharecroppers and planters, which she would depict in her novel *Boot-heel Doctor.* Cook wrote to Snow on March 16, asking him to arrange for her to stay with Dr. and Mrs. Albert Martin of East Prairie. Three days later, Snow responded to her letter, suggesting that she contact Mr. and Mrs. Ernest G. Gilmore of East Prairie. In his reply, Snow described Gilmore as a "natural born politician," who could place her in contact with "everything and everybody better than anyone I know." Of course, Snow added, she might want to visit Dr. Martin. But Gilmore, a farmer and county official, was associated with the local Social Security office, and that office certainly provided the best connection to "our lowly croppers."[70] Cook accepted Snow's recommendation and contacted the Gilmores.

On April 17, 1939, she wrote a long letter to her friend Edna Gellhorn, reporting on a three-day visit to "the sharecropper country." What she witnessed was a nightmare:

> I saw hungry women with scrawny babies gnawing at their breasts. I saw families where no child had a change of clothing. I saw families living under torn tents on ditch dumps—fine-looking people keeping alive on $3 a MONTH for the three or four of them. These were NOT sharecroppers who had taken part in the roadside demonstrations, but exfarmers, white people.[71]

68. Mitchell, "Homeless," 9.
69. *Sikeston (Mo.) Standard*, February 17, 1939.
70. Fannie Cook to Thad Snow, March 16, 1939, Fannie Cook Papers, box 5, folder 2; Thad Snow to Fannie Cook, March 19, 1939, Cook Papers, box 5, folder 2.
71. Cook to Gellhorn, April 17, 1939.

Reporting on the plight of the demonstrators, whom she called "Whitfieldians," she said they had been "dumped" by state police, who cleared their tents from the roadside. In Charleston, they lived in stalls made of cardboard boxes propped up against bare walls. They had no running water or privies. Others lived out in the country in similar circumstances. Women gave birth without medical care. "Sick, bilious people [lay] on the bed in another camp half-conscious with no one to find out whether they have the flue [*sic*] or smallpox!"[72]

One week later, she wrote Snow thanking him for introducing her to people in the Bootheel. Clearly she had observed both the middle and upper classes as well as the farmworkers. She had spent time with a physician, talked with a cotton ginner, "attended 3 Negro churches," and "in general found just the material I sought."[73] Cook's letter of April 24 was cheery and social in nature, in stark contrast to the letter to Gellhorn.

During her visit to the Bootheel, she had spent much of her time with socially prominent people, including a man from Indiana who ran a lumber mill at East Prairie. Of this gentleman, Mr. Stormes, she wrote:

> The last gentleman is going to call upon you (I hope with the jolly Mrs. Stormes) and he will say that I invited him. After talking to him about everything from the pheasants he is raising to the Necessity for independence in the human critter, I felt you two would like each other. If you find I was mistaken, you live far enough apart not to be too much bothered by having met.[74]

Cook's Bootheel hostess, Maude Gilmore, wrote to her on April 27, 1939, inviting her to visit again. In this letter, she described her recent trip to Malden with the Study Club of Charleston. Of this outing, she wrote that "We were guests of the Malden Club at an old Colonial home. It is an interesting place to go. They have negro servants, old period furniture and such a large old mansion."[75] There was an ironic disjunction between the world of the Gilmores and

72. Ibid.
73. Fannie Cook to Thad Snow, April 24, 1939, Cook Papers, box 5, folder 2.
74. Ibid.
75. Maude Gilmore to Fannie Cook, April 27, 1939, Cook Papers, box 5, folder 2.

that of the sharecroppers. Snow and Whitfield provided Cook with links to both of them.

Cook befriended Whitfield's wife, Zella, in suburban St. Louis. In a letter to Cook, dated August 13, 1940, Zella described the birth of a new granddaughter. In spite of Cook's efforts, the hospital in Kirkwood refused to admit "those old no good share croppers from South East Mo.," and so Zella's daughter went to "a colored doctor." Zella closed the letter, "Thanking you again for your kindness to my family and myself as well as for all kindness and help you are giving to the suffering people in South East Mo."[76]

With thousands of sharecroppers still homeless, Owen and Zella went to Washington to testify before a congressional committee. President Roosevelt invited them both to the White House. Zella told Eleanor Roosevelt what it was like to be a sharecropper's wife. Owen explained to the president how landlords cheated tenants out of their parity payments.[77] The Roosevelts agreed that the situation was disgraceful, and plans for federal housing projects were going forward. But the sharecroppers' plight remained desperate.

The Missouri Committee for the Rehabilitation of the Sharecroppers helped Whitfield establish a refugee colony on a rugged ninety-acre tract in southwestern Butler County, fifteen miles southwest of Poplar Bluff, on the east side of the Little Black River. William R. Fischer, a Mississippi County tenant farmer, initially purchased the abandoned farm with help from the Federal Land Bank in the summer of 1939. In September 1940, the Missouri Committee purchased the property from Fischer.[78]

On June 18 and 19, 1939, thirty-eight families—some two hundred people—arrived at the camp and set up tents provided by the federal government. One month later, more than five hundred dispossessed farmworkers (ninety-seven black families and eight white families) were in residence at the new settlement, called the Sharecroppers' Camp, or Cropperville.[79] White residents of Butler County complained to the sheriff's office that no African Americans had

76. Zella Whitfield to Fannie Cook, August 13, 1940, Cook Papers, box 5, folder 5.

77. Belfrage, "Cotton-Patch Moses," 101.

78. Fannie Cook, "Cropperville Gets a School," Fannie Cook Papers, box 26, folder 8; Butler County deed book 216, Page 242.

79. *Poplar Bluff (Mo.) Daily American Republic,* June 19 and July 19, 1939.

lived in that rural area before and that no sanitary facilities existed on the tract. Despite these protests, the refugees began creating a community.

Shirley Whitfield Farmer remembered with awe how the homesteaders cleared the land in the camp to lay out roads and build houses. As a young girl, she watched as "they took some logs, cut down the trees and took some heavy logs and then laid some heavy bricks on top of the logs, and that's how they started clearing away the heavy weeds and dirt, so they could clear the way to make a road." She also recalled that the houses were made of logs—"logs packed with mud. They didn't have a floor; they had dirt. The ground was the floor. But these ladies kept it so clean, you would have thought, like somebody living on a carpet."[80] When a new family arrived, everyone pitched in to build them a house. They also built a church, and a group of Quakers constructed a clinic.

The local school board and the county school superintendent sought help from the state, declaring that local schools could not accommodate the children of Cropperville.[81] When state officials refused to assign a teacher without a school building, Whitfield announced that he would construct one. His union friends sent money and lumber. The Committee for the Rehabilitation of the Sharecroppers purchased textbooks, and a reluctant local school board paid for equipment. A goat supplied milk, and the FSA provided hot lunches. Fannie Cook sent crayons and drawing paper, and the state provided a teacher.[82]

In the rugged Ozarks, lacking the rich Bootheel soil, the Cropperville community struggled to survive. Whitfield did some truck farming, worked part time in a lumber mill, and raised sorghum in cooperation with his neighbors. His daughter Shirley vividly remembered the process of turning sorghum into molasses. Without the necessary tools, the croppers had to "make their own machetes out of wood" to strip the leaves off the plants. They used mules and horses to operate the press that squeezed the juice out of the sorghum. Then they had to cook it and skim it to make the molasses. Watching

80. Farmer interview, April 7, 2002.
81. *Poplar Bluff (Mo.) Daily American Republic,* July 19, 1939.
82. Cook, "Cropperville Gets a School," 5–7.

this as a child, she wondered how the adults around her had learned these skills. She also learned the value of working together as a team; it was a hard life, she said, "a crazy life."[83]

In the hills of Butler County, unemployment was a constant problem. An observer from the American Friends Service Committee, a Quaker group that supported the colony, wrote about factionalism and tension in the community. Writing to Cook in St. Louis from the Friends Work Camp at Harviell, in the summer of 1942, Clarence H. Yarrow reported that

> people are discouraged and tired. . . . [Their] hopes have been raised and cast down. They don't know what to hope now. They take the attitude of "what's the use." They are looking forward to another bleak winter, scratching the hills for fuel and wondering how they can get work. There are not many employed now. They have nothing to do, but sit around and intrigue and get on each other's nerves.[84]

During World War II, some of the residents found new opportunities. Young people moved away, and the population dwindled. By 1943, many of the able-bodied men were gone.[85]

While Whitfield toiled in Cropperville, Snow mended his strained relationship with other Bootheel landowners and worked with them to obtain state and federal assistance for farmworkers. After the demonstrators were gone and emotions settled down, other planters joined Snow in expressing sympathy for the farmworkers.

---

83. Belfrage, "Cotton-Patch Moses," 102; Farmer interview, April 7, 2002. The community struggled for many years. In April 1964, the Missouri Committee for the Rehabilitation of the Sharecroppers conveyed a portion of the property (38.3) acres to the SEMO Baptist Fellowship (Butler County deed book 405, p. 498). In 1957, Zella Whitfield signed a warranty deed on behalf of the committee, conveying the remainder of the property to the SEMO Baptist Fellowship (deed book 546, p. 355). On August 29, 2001, the court awarded Friendship Missionary Baptist Consolidated District Association and the SEMO Baptist Fellowship full title to the entire tract (deed book 923, p. 456). A gable-front log house covered with shingles—which may have been Whitfield's house—remains on the property.

84. Clarence H. Yarrow to Mrs. Fannie Cook, July 8, 1942, Cook Papers, box 7, folder 4.

85. Farmer interview, April 7, 2002; Robert H. Forister, *Complete History of Butler County, Missouri*, 170–71.

In December 1939, nearly a year after the roadside demonstrations, sixty-five landowners met in Charleston to discuss the problem. Some planters who had adamantly opposed the strike now balked at the prospect of additional mass evictions, not only of black share-croppers, but of white tenant farmers, because of unfair administration of the labor provisions of the cotton control laws. Unanimously, the landowners adopted a resolution stating that cotton laborers were entitled to share in government subsidies.[86]

Fearing social instability, the landowners urged the U.S. Congress to enact legislation making it impossible for a landowner or a tenant to receive a sharecropper's portion of the subsidy payment. Speakers at the meeting, including Snow, reported that abuse of the subsidy system had resulted in large profits for landowners. This in turn led to the sale of large tracts of high-yield cotton land and the subsequent displacement of tenant farmers. In order to prevent further suffering, the landowners demanded legal reform and honest administration of the farm laws.[87]

Snow went to Washington to press the Department of Agriculture for changes in the laws governing AAA cotton benefit payments. At his urging, the department drafted amendments that would prevent landowners from turning sharecroppers into day laborers and pocketing the portion of benefit payments that the croppers were supposed to receive.[88] This was just what the woman named Annie accused Snow of doing, and it certainly was the reason why many croppers joined the roadside demonstrations. If Snow in fact did treat Annie so shabbily, he tried in the years following the strike to make things better for the croppers.

Early in 1940, Snow, along with Judge W. J. Melton, Miles T. Lee, and J. O. Bebout, represented Mississippi County at the Land-owners and Sharecroppers Conference in St. Louis. The purpose of the conference was to arrange for the Missouri State Employ-ment Service to help sharecroppers find jobs. In February 1940, the Landowner-Tenant-Sharecropper organization proposed that the FSA assist sharecroppers by expanding the labor rehabilitation program,

86. *Sikeston (Mo.) Standard*, December 15, 1939.

87. Ibid.

88. Article by Bruce Catton for the *Washington Constitution-Tribune*, reprinted in the Chillicothe, Missouri, *Tribune*, April 25, 1940, filed in Stark Papers, folder 1954.

helping farmers to relocate, financing new farms, and building homes on FSA tracts.[89]

In cooperation with the state of Missouri, the federal government developed and implemented several programs to aid farmworkers. Under the labor rehabilitation program, the FSA accepted responsibility for more than fifteen hundred families, who were on relief or eligible for relief. The FSA provided small grants to buy garden seed, tools, and other items necessary for subsistence. Many families also obtained loans to buy cows. In return for these gifts, grants, and loans, the families agreed to maintain and improve the properties on which they lived.[90]

The federal government constructed housing for workers on private land donated by landowners. Snow donated land to the Southeastern Missouri Scattered Labor Homes Program. Under this program, landlords leased three-acre tracts rent-free, and the government assisted with the cost—five hundred dollars—of building small houses for eligible families. At the end of ten years, the house and land reverted to the landlord.[91] Under this program, Snow leased two three-acre tracts to James M. and Helen Manker and Charles and Florence Manker.[92] James Manker would eventually become a partner of Snow's son-in-law, when he took over the farm.[93]

For families who could not obtain a lease on private land, the FSA constructed workers' homes on land purchased by the government.[94] In the Bootheel, construction on the Delmo Project houses began in 1940 and was completed in 1941. About six hundred houses were built in ten racially segregated villages of thirty to eighty-five houses each. Each village had a circular plan, with dwellings arranged in a loop around a central common area. Circle City in Stoddard County took its name from this design. The common area in each village included a well and water tower, community building, clinic, showers, laundry facilities, and demonstration kitchen.[95]

89. *Charleston (Mo.) Democrat*, January 18 and February 22, 1940.

90. U.S. Farm Security Administration, "Southeast Missouri: A Laboratory for the Cotton South," December 30, 1940, pp. 5–6.

91. Ibid., 7.

92. Deed book 103, p. 324; deed book 117, p. 336.

93. Corse interview, May 28, 2002.

94. FSA, "Southeast Missouri," 7.

95. Mitchell, "Homeless," 9. "Delmo" refers to the delta of Missouri.

At a cost of eight hundred dollars for each house, the government built four-room weather-tight residences with closets, built-in storage cabinets, and electric lights. Each house was complete with bedroom furniture, a cook stove, a coal heating stove, a dining table, and its own privy. Each home was built on four-tenths of an acre, enough room for vegetable gardens and small animals. In addition, the government leased farmlands near the communities.[96]

During World War II, labor became scarce, and landowners pressured residents of the Delmo villages to work for them as day laborers. Farmworkers in these communities had enough economic security to place them in a strong bargaining position with employers. In Mississippi County, landlords made a deal with state and federal officials to purchase the houses in a federal village just outside Wyatt and scatter the dwellings throughout the county. This action prompted David Burgess, a social activist from St. Louis, to comment, "When they [the landowners] didn't need them [the farm laborers], they cast them off, now when they need them they're buying them out."[97]

At the end of the war, the federal government reorganized the FSA as the Farmers Home Administration (FHA) and put the Delmo homes up for sale. In 1945, an interracial St. Louis committee, chartered as the Delmo Housing Corporation, purchased the remaining nine villages from the FSA. Support for this nonprofit organization came from the Washington University School of Social Work, the American Civil Liberties Union, and various churches. Setting the price of each house at eight hundred dollars, the original cost of construction, the corporation allowed residents to buy their houses, with a one-hundred-dollar down payment, making monthly mortgage payments of about ten dollars. By 1954, all the mortgages were retired, and the Delmo agency used the surplus funds to continue serving residents of the Bootheel with social, health, and educational programs.[98]

Through changing times, Whitfield continued to work for social justice. For several years, he traveled around the country, organizing

96. Alex Cooper interview with David Whitman, July 27, 1994, Portageville, Mo., Bootheel Project, audiocassette 15, Western Historical Manuscript Collection, University of Missouri–Columbia.
97. Alex Cooper, telephone interview with the author, June 1, 2001.
98. Ibid.

for the Food and Tobacco Workers. He also remained a popular preacher, often taking hundred-mile bus rides to give a Sunday sermon to a congregation of cotton workers. On his travels, he sometimes stopped at Snow's Corner, and from time to time Snow dropped in unexpectedly at Whitfield's house in Cropperville. Whitfield's daughter Shirley remembered seeing Snow talking in the yard with her father. He would sit with Owen and Zella and reminisce about the great roadside strike and leave them a gift of a brace of ducks.[99]

By 1950, Whitfield had left Cropperville to become pastor of a church in Du Quoin, Illinois. His daughter Shirley remembered attending the high school there, which was not segregated although it struggled with racial problems. Whitfield addressed school assemblies on the issue of prejudice. In 1955, he moved to Cape Girardeau to become a minister at the Second Baptist Church. At the time of his death in 1965, he was pastor of the Pilgrims Rest Baptist Church in Mounds, Illinois.[100] In the last years of his life, he became less active in the labor movement, but he continued to speak to his congregations about justice and racial equality.

99. Belfrage, "Cotton-Patch Moses," 102-3; Farmer interview, April 7, 2002.
100. Christensen et al., *Dictionary of Missouri Biography*, 793.

Bootheel Planter 7

Unless we miss our guess thumbs will be turned down on Thad Snow by landowners in Southeast Missouri in the future should he attempt to lead them as he has in the past. He knew what was brewing and helped to set the pot to boiling and instead of him doing a neighborly act and telling the landowners what was going on, stood with the negro Whitfield in getting the roadside setters going.

—*Sikeston Standard,* January 17, 1939

Snow betrayed members of his own class and stood with aggrieved farm laborers in 1939. The editor of the Sikeston newspaper confessed a complete failure to understand why Snow, a wealthy planter, "was so interested in the poor downtrodden tenant farmer."[1] Why would a capitalist betray his own self-interest, and, as the paper implied, that of his well-to-do neighbors, to side with the working class? Why would a white man sympathize with a black man's cause? Despite the editor's obvious animosity, he posed important questions, which deserve serious answers.

Two white middle-class contemporaries wrote books about the 1939 sharecroppers' strike. One, of course, was Snow; the other was Fannie Cook, the St. Louis novelist and activist. Cook and Snow were acquaintances. In the aftermath of the strike, he had helped introduce her to people in southeastern Missouri so that she could

1. *Sikeston (Mo.) Standard,* January 24, 1939.

prepare for her novel *Boot-heel Doctor;* she knew about his associa-tion with Whitfield, as well.

Although it was a product of Cook's imagination, the novel was based on observation and contained valuable insights on the situa-tion in the Bootheel. Its principal character was a white physician who befriended the leader of the sharecroppers' protest in 1939. Whitfield provided the real-life model for the fictional character Reuben Fielding. Joel Gregory, the white doctor, was a composite character, but in some ways he resembled Thad Snow.

The novel focused on Gregory as he visited sharecroppers' cab-ins, delivered their children, tended their ailments, and commiser-ated with them through the winter flood of 1937 and the uprising of January 1939. Cook, a physician's wife, was familiar with a doctor's routine. The local model for the physician may have been East Prairie's Dr. Albert Martin. It seems likely that she chose a doctor, rather than a planter, as a less compromised and more sympathetic hero. During the course of the novel, the doctor became more and more closely associated with Fielding, the black labor leader.

Like Snow, Gregory had paternalistic impulses. Although he had good intentions and was not an overt racist, he clearly felt superior to his poor white and black patients, treating them in a fatherly way, talking to them as though they were children, and feeling frustrated by their refusal to follow his advice consistently. At the climax of the novel, however, he recognized the strength and dignity of the poor white and black sharecroppers and especially of their black leader. By going out on strike against the big planters, he saw that "By God, they're about to do the bravest thing's ever been done in the boot-heel."[2] These attitudes closely paralleled Snow's ambivalent elitism, his growing sympathy with agricultural workers, and his dramatic realization that the people he regarded as lowly croppers had done something of unquestionable historical significance.

These two fictional characters, one white and one black, make crucial choices that place them at the center of a changing social world. During the course of the novel, Fielding develops from a careful, observant black man shrewdly surviving in white society, to the bold leader of a protest movement. In an early chapter, he works quietly at the Red Cross shelter, performing menial tasks for

2. Fannie Cook, *Boot-heel Doctor,* 235.

no pay, helping black and white families devastated by a flood. In the aftermath of that natural disaster, he begins to appreciate the extent of the sharecroppers' misery and the relentlessness of the planters' greed.[3]

Cook described the influx of Red Cross workers, volunteers, and government functionaries into the Bootheel during the 1937 flood. These outsiders upset the equilibrium of her fictional community, North Cotton, as planters tried to maintain their dominance over tenants and sharecroppers. Public charity freed the farmworkers from total dependence on their landlords, who were no longer the sole providers of shelter, clothing, and sustenance. Planters tried and failed to control the cheerful, professional efforts of the Red Cross, and suddenly the time-honored bonds of personal fealty were broken. White croppers were torn between loyalty to their race and the realization that they were an underprivileged class. Black croppers challenged old assumptions about their proper place in the planters' world.

Cook portrayed Fielding as a smart, articulate man with a clear understanding of interracial dynamics. Working as a janitor in a Red Cross shelter, he keeps a sharp eye on the white volunteers who dole out food and clothing. As a minister, he urges his people to be calm and patient, but he also advises them to protect their own interests. To a white companion, he explains, "every time I go to letting my folks keep they eyes turned to Heaven, staring too hard at the angels, somebody come along and snatch they dinner off they plate."[4]

Displaced from their homes and thrown together in refugee camps, Cook's fictional sharecroppers develop a class-conscious appreciation of their plight:

> Waiting for the flood to recede, the peoples of the Bootheel were like a large family whose photograph has been snapped without warning. Astonished and hurt, they studied their portrait, fascinated by their own unsightliness.
>
> Social workers in raincoats had looked up from their papers to ask: "How many children?" "Are you a tenant?" "Or a share cropper?" "Have you enough money to rebuild?" "What did you lose?"[5]

3. Ibid., 40, 63.
4. Ibid., 105.
5. Ibid., 61.

Caught in the spotlight of public attention, the sharecroppers had a chance to look at their situation and realize that it might be a national, not just a local, problem.

As the story continues, Fielding develops a race and class consciousness that inspires him to take a dangerous stand. From the flood, he learned that disasters did not operate equally on whites and blacks, and government assistance could be appropriated by powerful white elites. Angered and activated, he preaches a sermon in a schoolhouse, admitting that the flood was bad, but "When trouble comes to black people, the worst kind's usually the kind Man made." Hearing this sermon, the white doctor is frightened and thinks, "Colored men had been lynched for saying less."[6]

While Fielding emerges as the leader of a protest movement, Gregory watches him with awe and more than a touch of envy. Reuben is making history, and the doctor is ashamed not to be part of that history, ashamed not to be making an equal sacrifice. The doctor receives a share of disapproval for inviting Fielding into his home and being seen with him in public. Hester, the doctor's wife, worries continually about his safety. But it is Reuben who takes the real risks. In the end, Cook's plot allows the doctor to become the hero: When a crowd attempts to lynch the black leader, the white healer comes to his rescue.[7]

Snow never became a hero in such an obvious, active way. He was brave in that he took an unpopular stand, but he never put his life on the line or rescued Whitfield from physical danger. Also, there were many ambiguities in Snow's position. He was not a doctor; he was a planter, and he directly profited from sharecroppers' labor. Two of his own farm laborers joined the roadside protest. His heroism grew from his having endured the wrath of former colleagues and the taunts and threats of a newspaper editor. After the protest, having mended his relationship with other planters, he worked doggedly, albeit undramatically, to win government aid for farmworkers. After the climax, the novel ended, but real life went on in the Bootheel.

Real life could be dramatic enough. Snow continued to study the farm-labor problem, while another white man, Wade Tucker, tried to keep unionism alive in the region. Tucker was a larger-than-life

6. Ibid., 102.
7. Ibid., 197, 211, 265.

Bootheel hero, who came to the region as a logger and won fame as a wrestler. In the 1920s, he raided bootleggers and moonshiners as sheriff of New Madrid County. Defeated for office in the 1930s, he returned to work as a tenant farmer.

Although he did not participate in the roadside strike, Tucker became involved in labor militancy in the spring of 1939. He traveled around the Bootheel, speaking to crowds of farmers as an organizer for the Associated Farm Laborers, Sharecroppers, and Tenant Farmers of Southeast Missouri (AFLST), a rival of the STFU. While the STFU organized workers against landowners, the AFLST tried to organize cotton farmers against ginners.[8]

The editor of the *Sikeston Standard* urged planters to mechanize their operations in response to Tucker's activities. In the May 19 issue, he encouraged producers to attend a demonstration of a cotton chopper at the Van Ausdall brothers' farm two miles west of Caruthersville. This might be an antidote to labor troubles, because "If Wade Tucker succeeds in organizing the farm labor in Southeast Missouri and places them on eight hour shifts at $2 per day, this cotton chopper will do the work and no strikes."[9]

In July 1940, Tucker made a speech from his automobile at Peach Orchard (Pemiscot County), Missouri. After he finished and was shaking hands with the crowd, an angry man confronted him. Tucker responded with a joke, and the man shot him twice in the abdomen. At the hospital in Poplar Bluff, Tucker suffered through two operations and a bout with pneumonia—but he survived. Neighbors rallied around him, and friends helped him get a job as a supervisor for the Farm Security Administration.[10]

Hearing about the shooting, Cook contacted Snow, asking for information about Tucker. Snow, who was leaving soon on a trip to Mexico, replied that he had talked with Tucker's physician at the hospital in Poplar Bluff but had not been able to visit Tucker. With friends in the Bootheel, Snow had helped to raise money for Tucker and his family.[11]

On the subject of the attack, Snow said he had heard conflicting reports. He believed that Tucker had said something insulting to the man who confronted him, and

8. *Sikeston (Mo.) Standard,* May 12 and 23, 1939.
9. Ibid., May 19, 1939.
10. *St. Louis Post-Dispatch,* July 12, 1941.
11. Thad Snow to Fannie Cook, July 29, 1940, Cook Papers, box 5, folder 5.

without a doubt the motive for the shooting rose out of righteous zeal to preserve the social order. Wade, of course was quite unarmed and unaggressive. The best I can get it indicates that he was chaffing [*sic*] the zealot, which, you know, is an outrageous thing to do. But I like to do it, and if it really isn't a crime that merits shooting, I'd like to see you establish the fact so I won't have to cramp my style.[12]

Clearly, he viewed Tucker as a kindred spirit, and he also regarded Cook as a comrade in the fight against "the zealot." In a conspiratorial tone, he asked in his letter, "Aren't we wild and wooly?"

For a time in the 1940s, Snow and Cook became literary rivals, both trying to publish books on the sharecroppers' roadside demonstration. In early 1941, after Houghton-Mifflin rejected the manuscript of *Boot-heel Doctor*, Cook wrote furiously to her literary agent, insisting that her work had to appear in print before Snow's. With obvious malice, she asserted that much of his work was being "ghost-written." In March 1941, she wrote to her agent about an "intimate chat" with Snow, in which he apparently expressed admiration for Lawrence Dennis's last book. Because of this, she feared that Snow would embrace fascism, "thus in his aged innocence lighting tinder he'll never live long enough to extinguish. Another reason my book ought to come out first!"[13]

Her hostility toward fascism was implacable and stemmed from her identity as an American Jew. Born Fannie Frank, she was the daughter of German Jewish immigrants. Her father was a partner in Fishlowitz and Frank, a large neckwear company in St. Louis.[14] In 1915, she married Dr. Jerome S. Cook, a practicing physician, who became director of medicine and chief of staff at Jewish Hospital.[15] In the 1940s, she published several novels treating the issue of justice for all racial and ethnic groups, including the Jews.

---

12. Ibid.

13. Fannie Cook to "Margot" [literary agent], January 6, 1941, Cook Papers, box 12, folder 18; Fannie Cook to "Margot," March 8, 1941, Cook Papers, box 12, folder 18. Lawrence Dennis, an American fascist, wrote *The Dynamics of War and Revolution.*

14. Helen Frances Levin Goldman, "Parallel Portraits: An Exploration of Racial Issues in the Art and Activism of Fannie Frank Cook," 2.

15. Jean Douglas Streeter, Fannie Cook Papers, Register (1988), Missouri Historical Society, St. Louis, 1988, p. 3.

Although her social convictions were genuine and admirable, her insinuation about Snow's latent fascism was grossly unfair. In a 1937 letter to the editor of the *St. Louis Post-Dispatch*, Snow admitted reading a fascist newspaper, the *Deutsche Weckruf*, in addition to the *Saturday Evening Post*. He even confessed to reading the fascist bible, Adolf Hitler's *Mein Kampf*, which he found to be "overpowering." Some readers may have interpreted this comment as an endorsement of Nazism, but Snow went on to remark sarcastically that he wondered if he would make a good fascist. First, he said, he would have to find out if he was pure Aryan. He wondered if there was a medical test for that. If he failed the test, he supposed he "would have no right to feel with Herr Hitler" that he was chosen to do the Lord's work of exterminating the Jews. Secondly, he was not sure that the Jews in his community were like the ones in Germany, who, according to Hitler, were "international" Jews, and, apparently, all communists. His tone was flippant, but his message was a serious one. He saw clearly that Hitler's philosophy was not only dangerous, but also absurd.[16]

Snow liked Cook and certainly did not deserve her accusations. But he may have shared in a local belief that she, as an outsider, had no right to criticize the Bootheel. In March 1941, the *Post-Dispatch* printed an editorial by Charleston editor Art Wallhausen, castigating Cook for poking her nose where it did not belong. He said she knew nothing about society in Southeast Missouri. Furthermore, he argued, she ought to be tending her own garden in St. Louis, where homeless derelicts accosted every visitor from out of town.[17] Snow, as it turned out, had read and approved of the piece before Wallhausen sent it to the St. Louis paper.[18]

Snow and Wallhausen were friends in spite of their political differences. Years later, Wallhausen's wife, Millie, recalled that "Thad was an eccentric. He was a very odd person. My husband—he and my husband were intellectual debaters. They just loved each other for that reason."[19] Wallhausen sympathized with the sharecroppers

16. *St. Louis Post-Dispatch*, July 19, 1937.
17. "Editor Wallhausen Hauls Off," clipping from *St. Louis Post-Dispatch*, March 7, 1941, in Cook Papers, scrapbook.
18. Thad Snow to Ralph Coghlan, n.d., Cook Papers, scrapbook.
19. Mildred "Millie" Wallhausen, interview with Ray Brassieur, Charleston, Mo., June 14, 1996, Bootheel Project, collection 3965, audiocassette 2, Western Historical Manuscript Collection, University of Missouri–Columbia.

but believed the landlords were being unfairly characterized as mean and grasping. In his view, Cook had contributed to the character assassination of the Bootheel elite.

The Wallhausen incident revealed Snow's ambiguous position in, and feelings about, the Bootheel. Although he sometimes cultivated his image as an outcast, he obviously maintained friendships with local dignitaries, including Wallhausen. While he did not openly approve of Wallhausen's tirade, he wrote to Ralph Coghlan, editor of the *Post-Dispatch:* "Dear Ralph, My righteous indignation over your printing Art's piece is somewhat modified by the fact that Art brought it over for me to read last Sunday afternoon. We went over it together. I even made some suggestions. Art thought the use of my name might help sell it to you. I was flattered, of course, so offered no objection."[20]

Coghlan sent the note on to Cook, who replied with unctuous good humor, "It was good of you to send me Thad's note. He's always the mischievous, adorable boy at heart—eh, what?"[21] This was quite affectionate for a woman who feared Snow was being seduced by fascism. Possibly this was her way of maintaining a cordial relationship with Coghlan, while dismissing Snow as a capricious, willful child who should not be taken seriously.

While Snow did not measure up to the image of Dr. Gregory as a local hero, he was more than a "mischievous boy" just trying to stir up trouble. He continued his association with Whitfield in a racially charged atmosphere that made the Bootheel an uncomfortable, sometimes dangerous, place for black people and the white people who associated with them. In 1942, one of the novel's events took on a terrible meaning when a white mob lynched a black man in Sikeston.

Snow took no action and made no formal public statements. His daughter Fannie was away at college in Tennessee, but she remembered her father's "anguish over it all. He just thought it was terrible." The lynching, on Sunday, January 25, inspired widespread outrage, prompted an investigation by the attorney general of the United States, and helped focus federal attention on racism.[22]

The ugly events began shortly after midnight, in the Sunset Addition, when citizens and lawmen rushed to the aid of a white woman

20. Snow to Coghlan, n.d.
21. Fannie Cook to Ralph Coghlan, March 13, 1941, Cook Papers, scrapbook.
22. Delaney interview, March 19, 2001; Capeci, *Lynching,* 50–51.

bleeding from stab wounds. Police officers apprehended a black man named Cleo Wright, who lashed out at them with a knife. Later that morning, word spread that Wright had confessed to the crime. A mob gathered around City Hall. County Prosecutor David L. Blanton tried to defuse the situation, but he could not stop the crowd from grabbing Wright, kicking him, dragging him through Sunset Addition, dousing him with gasoline, and burning him to death.[23]

In the aftermath of this outrage, Charles Blanton, the Polecat, tried to defend the existing social order in the Sikeston paper. Not even his son, the prosecutor, could prevail upon him to maintain a dignified silence. In his newspaper columns, he lashed out at anyone who criticized this vigilante action, wondering, for instance, how Fannie Cook might feel if a black man climbed into her bed some night.[24] These were the same racist sentiments that prompted Blanton to question the origins of Snow's sympathy with the masses.

While he was not alone in his anguish over the lynching, Snow differed from most of his white neighbors in his feelings on issues of race and class. What made him so different? He hailed from the Midwest, rather than the South, but that was true of others. Caverno was a northerner, but he remained true to the landlord class during the roadside strike. Snow's family had Republican connections, and his father served in the Union army, but many Unionists, Republicans, and even former abolitionists espoused a racist philosophy during and after Reconstruction. While Snow could be paternalistic, he never embraced the idea of white superiority.

As Snow identified more and more with the sharecroppers' cause, his friends and neighbors increasingly labeled him a misfit. Some of his old friends deserted him. Others, like Millie Wallhausen, saw both good and bad qualities in him. Years later, she remembered that he was "very good to the people who worked for him" and also "a good friend." On the other hand, she said, "You better not cross him. He could cut you to ribbons with his words."[25] Snow could be a compassionate man, but he could also be insensitive to other people's feelings. These two conflicting qualities made it possible for him to sympathize with the sharecroppers and also to stand alone against his white neighbors.

23. Capeci, *Lynching*, 13–23.
24. Ibid., 155.
25. Wallhausen interview, October 28, 2002.

He tended toward idealism, which blinded him at times to the nuances of human experience. With his second wife, Lila, he shared an intense intellectual comradeship. After her death, he transferred this psychological bond to his youngest daughter, Emily, a beautiful girl who inherited her parents' passion for political philosophy. At the same time, he became estranged from his only son, Hal, and failed to comprehend problems in Priscilla's marriage. Priscilla's friend Millie said Snow doted on Emily and paid little attention to any of his other children, including Fannie, who could not match her sister's beauty or intellectual passion. In his memoirs he wrote about traveling and discussing ideas with Emily, scarcely mentioning her siblings.

Some of his ideas came from books. But in a much more intimate and complex way, his ideas emerged from the experience of living in the Bootheel. Near the end of his life, he wrote: "I've seen things happen, and things have happened to me, in Swampeast Missouri that I could not possibly have experienced anywhere else in the Midwest, or in this whole United States, for that matter."[26] Among these things were the quick and brutal closing of a frontier, the complete alteration of the local environment, four disastrous floods, the exploitation of farmworkers, a relentless economic depression, and a labor uprising led by a member of an oppressed racial minority.

The title character in *Boot-heel Doctor* had a kind of conversion experience that was unlikely to happen in real life. For the fictional doctor, the epiphany came when he invited a black man, Reuben Fielding, to come into his home through the front door. In the real world such moments are less dramatic and more diffuse, but clearly, as Snow developed a relationship with Whitfield, he moved out of the closed circle of former white friends and associates of the planter class and into new intellectual and social territory.

Although Snow was unusual, he was not the only Cotton Belt paternalist to espouse liberal views in the first half of the twentieth century. Historian Fred Hobson noted the appearance of the "racial conversion narrative" as a new subgenre of southern literature in the 1940s. In these narratives, white southerners who had benefited from the suppression of a racial minority confessed their sins, condemned slavery, accepted the justice of Confederate defeat in the Civil War, and recognized the evils of segregation. These were loyal

26. Snow, *From Missouri*, 2.

Southerners who hoped to find some form of redemption for their beloved land. Some placed their faith in the emerging civil rights movement and its black southern leaders, viewing the black southerner as more courageous, wiser, and more righteous than the white majority.[27]

Snow's insights resembled, but did not duplicate, those of his literary counterparts in the Deep South. He felt the same guilt for having profited from the exploitation of a racial minority. Whitfield's intelligence, bravery, and dignity convinced him that the victims of oppression were the ones who possessed the moral authority to lead the white majority in a new and better direction. For most white southerners, these perceptions led to a comfortable position within the liberal mainstream of American politics. For Snow, the implications were more radical. In the dark face of oppression, he saw the possibility of a worldwide transformation, foreshadowed in the triumph of Mahatma Gandhi's peaceful overthrow of the colonial rulers of India.

Snow did not consider himself a radical. To him, a radical was someone who thought he had all the answers, and he never made that claim. In general, he resisted the urge to affiliate with any radical group. He considered himself merely "irregular,"[28] although he was deeply concerned about the philosophical foundations upon which the American political system rested. He did not become a communist, although some people accused him of that; he did not become a fascist, although at least one person accused him of that. Regardless, he definitely strayed from the mainstream of American political and social thought.

When liberals and conservatives embraced militarism after Pearl Harbor, he vehemently opposed America's participation in World War II. As his daughter Fannie recalled, "He cursed the war. He was just rabid over it." Pride gleamed in her eyes as she quoted her father: " 'At least it got us out of the Depression. That's why we got into the war.' That's what he said." In a strong voice, full of conviction, she asserted, "I believe that."[29] Long after his death, Fannie remained loyal to the vision that set him apart from most of his peers.

27. Fred Hobson, *But Now I See: The White Southern Racial Conversion Narrative*, 1–2.
28. Snow, *From Missouri*, 335.
29. Delaney interview, March 19, 2001.

Missouri Pacifist 8

> I think I regard Mahatma Gandhi as the greatest teacher of my lifetime.
> A few years back I was so naive as to believe that some of his teaching
> was sinking in a bit, here and there in the world. As I saw it, his finest
> idea was: You don't achieve good and lasting ends by evil means. This
> still seems to me to be a splendid idea.
>
> —Thad Snow, *From Missouri,* 341

Snow stood with a tiny minority against the United States' belligerence in World War II, even after the Japanese bombed Pearl Harbor. A few days after the attack, public opinion polls found that 96 percent of the nation approved of Congress's declaration of war. Representative Jeannette Rankin, who had previously voted against American participation in World War I, voted "no" again on the Monday after Pearl Harbor. But times had changed. During the First World War, Congress had debated the issue passionately, and fifty members of the House voted "no." In 1941, there was only one "no" vote. "This time," wrote Rankin, "I stood alone."[1]

Not quite alone. In a farmhouse in Southeast Missouri, Snow and his daughter Emily had long, serious talks about the morality of war. During their travels in Texas, they lodged near an aviation training field. Quite frequently they had heard of young trainees there suffering injury or death. Emily wondered if the sacrifice was worth-

---

1. Lawrence S. Wittner, *Rebels against War: The American Peace Movement, 1941–1960,* 35.

while, and Snow tried to explain to her why it was necessary to be ready at any time "to fight to protect our democratic institutions." According to Snow, that answer failed to satisfy his daughter, because, "She says it is too high sounding to mean much, and reminds me that I have said we aren't aiming to be pulled into another foreign war to end war and to save democracy. She also reminds me of some remarks I have made about other ways to save democracy besides going to war over it."[2]

Although Snow condemned Japanese aggression, he looked back on the history of aggressive actions by his own country. For example, he believed Americans had conveniently forgotten the empire-building war against Mexico in the 1840s. If at that time some other nation, such as Japan, had demanded that the United States pull out and return all conquered territory to the Mexicans, "we would have been shocked and even angered" by such interference in our country's affairs.[3] But American aggression was just as immoral as that of any other nation.

These were not entirely new ideas. In the 1840s, the Mexican-American War inspired an indignant protest, beautifully articulated by Henry David Thoreau. The New England philosopher spent a night in the Concord, Massachusetts, jail for refusing to pay his taxes as an act of passive resistance against America's military action. By refusing to pay his taxes, he withdrew his support from a state engaged in an unjust adventure. He also expressed his moral revulsion against slavery by renouncing the authority of a nation that bought and sold human beings on the steps of its Capitol. His brief incarceration did not prevent the war or stop the sale of slaves, but his essay, *Civil Disobedience*, published in 1849, developed the idea of nonviolent resistance to injustice, which eventually had a profound effect on the thought of Mahatma Gandhi and Dr. Martin Luther King Jr.

Snow read widely in the literature of pacifism, including the works of Gandhi and Leo Tolstoy. Count Tolstoy, author of the monumental novel *War and Peace*, asserted that the Christian ideal of universal brotherhood prohibited killing even in response to provocation. In his philosophy, there was no such thing as a just war. The institutions

2. Essay opposing rearmament, Snow Papers, folder 54.
3. Thad Snow, "Reflections on War Hysteria," Snow Papers, folder 52.

of government and private property were inherently unjust and anti-Christian. In Tolstoy's view, the only protection against violence was the majestic power of Christian love. Although Snow was not a churchgoer, he agreed with Tolstoy that war had complex causes that could not be addressed by military action. What the world needed was a change of mind and spirit.[4]

Tolstoy's ideas influenced other American pacifists. William Jennings Bryan, the Nebraska populist, embraced Tolstoy at the turn of the century. During the winter of 1902–1903, Bryan visited the author at his home in Russia. For the next decade, the American politician extolled nonviolence as the proper course of action both for individuals and states.

A religious fundamentalist, Bryan warmly received the vision of universal brotherhood and the gospel of love, although he could not accept Tolstoy's rejection of any form of government as evil. In 1910, he declared: "I believe that this nation could stand before the world today and tell the world that it did not believe in war, that it did not believe that it was the right way to settle disputes, that it had no disputes that it was not willing to submit to the judgment of the world."[5] When Bryan became secretary of state in 1913, he was the first avowed pacifist to take responsibility for the nation's foreign policy. He resigned from his post in June 1915 because of his opposition to the United States' impending involvement in the First World War.

In that momentous year, 1915, Gandhi began a campaign of nonviolent resistance that would eventually free India from British domination. As a young lawyer, he had already led a moderately successful campaign to win civil rights for the Indian minority in South Africa, where he spent twenty-one years of his life. In 1907, while in South Africa, he read Thoreau's essay on civil disobedience, which provided an excellent rationale for actions he had already undertaken. Tolstoy's writings also inspired him, deepening his Hindu beliefs in asceticism and noninjury. Until his assassination in 1948, he practiced and affirmed the universal effectiveness of the nonviolent method of conflict resolution.[6]

4. Thad Snow, "Tolstoy's *War and Peace* and Roosevelt's Policies," [May 1941?], Snow Papers, folder 58.
5. Peter Brock, *Pacifism in the United States, from the Colonial Era to the First World War*, 935.
6. Peter Brock, *Twentieth-Century Pacifism*, 68–92.

In America, the Socialist party rallied against U.S. participation in the First World War. Eugene V. Debs, the Indiana labor leader and Socialist presidential candidate, went to prison for violating wartime restrictions on free speech.[7] In Missouri, Kate Richards O'Hare presided as chairwoman of the Socialist party's War and Militarism Committee at the St. Louis Emergency Convention in April 1917. During the summer of that year, she toured the country, giving a series of antiwar speeches that resulted in her arrest and prosecution under the Espionage Act. She served three years of her five-year sentence in the Missouri State Penitentiary, denouncing prison conditions in letters to her husband, Frank. In 1920, President Woodrow Wilson commuted her sentence. Two years later, she organized a crusade to free more than one hundred war protesters who were still serving time in prison.[8]

In the 1940s, after Kate divorced Frank O'Hare, he and Snow became friends. Like Snow, O'Hare contributed frequent essays to the *Post-Dispatch*. Friends and acquaintances flocked to his home in St. Louis to discuss the international crisis. Despite his impassioned objections to American participation in World War I, he changed his mind after Pearl Harbor. To some extent he felt relieved to find himself in the mainstream of public opinion. Snow could not join him there, but he and Emily paid visits to O'Hare during and after the war.[9]

Many other socialist leaders, including Norman Thomas, abandoned their pacifist ideals after Pearl Harbor. In 1940, the Socialist party condemned Hitler's war as "unholy" yet insisted on absolute American neutrality. But the party rapidly disintegrated, with many members calling for economic and military aid to England and the Allies. One group of disaffected socialists formed the Union for Democratic Action, which later became Americans for Democratic Action. Members of Thomas's own family disagreed with his antiwar stance. His youngest son joined the American Field Service a month before Pearl Harbor. After the Japanese attack, Thomas sadly concluded that there was no alternative to war.[10]

7. Nick Salvatore, *Eugene V. Debs: Citizen and Socialist*, 293–96.
8. Bonnie Stepenoff, "Mother and Teacher as Missouri State Penitentiary Inmates: Goldman and O'Hare, 1917–1920."
9. Peter H. Buckingham, *Rebel against Injustice: The Life of Frank P. O'Hare*, 192. In letters written in 1948, O'Hare described his warm feelings toward Snow and his daughter Emily. Frank O'Hare Papers, Missouri Historical Society, St. Louis.
10. Wittner, *Rebels against War*, 21, 37.

Pearl Harbor triggered an intense emotional reaction throughout America. Before the attack, public opinion remained divided on the need for intervention. Polls taken a few days after the bombing revealed nearly unanimous support for war. This was not a reasoned response, but an expression of outrage at a cowardly and deadly act. Snow did not share this reaction. "Pearl Harbor ought to simplify everything for me," he wrote, but in fact it made things more complicated. "[I] would like to believe that I'm the innocent victim of treacherous and wanton assault. But I am confused."[11]

In the Bootheel, giant headlines announced that Congress had declared war. Blanton of the *Sikeston Standard* yet again revealed his racism when he wrote, "At last the brown bellied Japs have struck."[12] Other editors were more refined, but still foursquare behind the war effort. A front-page editorial in the *Charleston (Mo.) Democrat* eloquently expressed the feelings that Americans almost unanimously shared. First, the editor commended England and the Allies, who had carried the burden of defending democracy. "It is now our turn," the editor wrote, "to put our shoulders to the wheel as one man to preserve our glorious heritage of freedom in a world gone mad." Feelings of nationalism and racism surfaced as the editor castigated the "kindled brute passions and grandiose schemes of the Japanese, German, and Italian murderers."[13]

Patriotism and altruism motivated many people to support the war effort. Young men from southeastern Missouri and southern Illinois crowded the navy and marine recruiting offices in Cape Girardeau.[14] An editorial in that city's newspaper noted that before December 7, enlistments had been declining. According to the editorial, "There can be no quibbling, no waiting for others to do it, every one of us must give as much as we can. Some people talk patriotism and do nothing else. Others practice patriotism by helping as much as they can. The opportunity to practice loyalty is now at hand."[15]

Mingled with the crusading passion that called men to preserve democracy was a vengeful, militaristic spirit. The editor of the

11. Thad Snow, "Reflections on War Hysteria," 1–2, Snow Papers, folder 52.
12. Capeci, *Lynching*, 13.
13. *Charleston (Mo.) Democrat*, December 11, 1941.
14. *Cape Girardeau Southeast Missourian*, December 9, 1941.
15. *Cape Girardeau Southeast Missourian*, December 15, 1941.

Charleston paper reflected that the treaty that ended World War I had previously seemed harsh and punitive. In light of more recent events, he concluded, however, that:

> Instead of the Treaty of Versailles being a cruel and inhuman shackle around the throats of the Hun oppressors it now is evident that the Treaty was not cruel enough. It should have stifled the chances of another band of madmen roaming the world with blood stained hands, and with fangs dripping blood, subscribing to the philosophy that "might makes right."[16]

In order for civilized people to go to war they had to become savages, or at least, they had to perceive their enemies as savages.

While he was in Washington lobbying Congress to change farm policy, Snow observed the country's rising war fever. In the spring and summer of 1939 and again in the spring of 1940, he made extended trips to the nation's capital, despite continuing problems with his legs. According to Washington correspondent Bruce Catton, Snow had "hard sledding," because "large bodies move slowly, government departments don't like to admit mistakes, and big cotton planters don't get listened to very eagerly down at Agriculture anyway."[17] Ultimately, his lobbying efforts failed, partly, he believed, because of his lack of experience and partly because war preparations had changed the political and economic climate.[18]

As early as 1937 or 1938, Snow began to fear that the Roosevelt administration might resort to war as a way to restore prosperity. By that time, after half a decade of New Deal programs, the economy had stalled again, and more than ten million Americans remained unemployed. Jobless people were becoming desperate. When Snow picked up hitchhikers and talked to them about the state of the world, he found them almost unanimously in favor of military action by the United States in the worldwide conflict. They had no ideological commitment to war, but they thought that war would make jobs for them. Snow thought so, too.[19] But he still hoped the

16. *Charleston (Mo.) Democrat*, December 11, 1941.

17. *Chillicothe [Mo.] Tribune*, April 25, 1940, clipping in Stark Papers, folder 1954.

18. Snow, *From Missouri*, 290–91.

19. Ibid., 337–39.

United States would not resort to armed force against Germany and Japan.

He could not accept war as a solution to economic stagnation. Although he sympathized with jobless men, he was frightened by the idea that war might be profitable or expedient. He knew that since World War I, the so-called civilized nations had developed new and ghastly methods of war making, but "had learned less than nothing about peacemaking." Once the new weapons were unleashed, no one would be able to control them. War, he believed, was a monstrous relic, "like a dinosaur out of the past making its uncharted way through a packed market place."[20]

Snow watched, with growing consternation, as his friends and associates gradually accepted and ultimately embraced the idea of war. As an increasingly disgruntled New Dealer, he felt betrayed by Democrats who fell under the spell of what he considered war hysteria. To him, it appeared that Americans in the political mainstream began to view war, not as a horror, or even as a grim necessity, but as a holy crusade. Somehow, in the minds of well-meaning people, "It was to be a kind of New Deal extension that would spread out all over the world, to the great material and spiritual benefit of all mankind."[21] Support for this idea rose and subsided throughout the late 1930s and into the 1940s.

Without considering the personal consequences of his actions, in private and in public, Snow condemned the war. His loyalty came into question when he asserted western powers had provoked the Japanese attack. Historically, in his view, the bombing resulted from centuries of colonialism. Racial passions flared when he insisted that the United States had cooperated openly and egregiously with England and France in an effort to dominate the "pigmented people of the East."[22]

Snow believed that England and its American allies had started the chain of events leading to Pearl Harbor by taking up the "white man's burden" and attempting to control the destiny of Asia. In this way, he reasoned, the war grew inevitably from a long history of imperialism, which had deep roots in Anglo-American assumptions of racial superiority. Thus, he concluded, the worldwide con-

---

20. Ibid., 337.
21. Ibid., 338.
22. Ibid., 4–5.

flict resulted from American racism as much as from Japanese nationalism or German anti-Semitism. There was, he realized, a connection between injustice in the Bootheel and war in Europe and Asia.

A long history of class distinctions, aristocratic privilege, injustice, and oppression had led inexorably to a series of military debacles around the world. Snow directly compared the first half of the twentieth century to the Napoleonic era. Like Napoleon, Hitler had set out on a crusade of conquest, but he was not the architect of that crusade. In the minds of many Americans, Hitler was a devil, a superhuman engine of evil. To Snow, however, he appeared as "too inevitable, a too explainable product of conditions to which my own country and its policies have contributed generously."[23]

Snow blamed the war not only on Hitler, but also on Roosevelt. In some of Snow's writings, he accused Roosevelt of manipulating events and using artifice to drag the United States into the conflict. But Snow concluded that Roosevelt was not evil. As an American leader, trying to deal with immense economic problems, he became part of events that were larger than any individual.

Ultimately, Snow believed that there were no devils and that no one man deserved blame for the war. By allowing injustice to continue in this country, by unquestioningly supporting England in its quarrels with other world powers, by pushing policy makers toward war, and by considering mobilization for war as a possible solution to domestic problems, Roosevelt became involved in a downward spiral of madness that could only end in cataclysmic violence. Questions of intention and responsibility were immensely complicated, in Snow's words, "The question of Free Will and Necessity being of course, one of the unfathomable mysteries of life. A still darker mystery it is in the life of nations, and mankind in general."[24]

In the 1940s, Snow grappled with questions about what had gone wrong with the world's political and economic systems that seemed to lead repeatedly to war. In a 1945 pamphlet, he wrote that the nations of the world needed to reexamine their ideas and institutions, especially in the economic realm. Why?

23. Snow, "Tolstoy's *War and Peace* and Roosevelt's Policies."
24. Ibid.

The old modes obviously had served us so poorly that the peoples of the world were unable to live at peace with themselves. Instead, all of us turned to murderous, suicidal, insane war. Something must have been wrong with our ideas, habits, and institutions else we should not have turned on ourselves to destroy ourselves and all our works in such manner.[25]

To him, the war was not a glorious crusade to save the world, but an admission of abject failure. More than one hundred and fifty years of economic development, resulting from the industrial revolution, had created a complicated world. In order to cope with new realities, nations needed to rethink their policies and reshape their economic, political, and social systems. Snow said the world needed "new institutional clothes all around." But restructuring a nation required intellectual courage and honesty. For many people, it was easier to blame their problems on other nations, other races, someone other than themselves, and to vent their frustrations in monstrous violence. As Snow expressed it, "Some say that the peoples of the world went to war because that was less strain on their minds than picking out and fitting up a new institutional suit. But it may be that the war will hasten, rather than retard, our need for change."[26] The war, he was sure, would not solve anything.

The Great Depression shattered his faith in free enterprise. When the war restored prosperity, he did not simply count his blessings. Where, he wondered, did all this new wealth come from? In 1945, he wrote, "Of course, the financing of the miraculous production of this war by ourselves, our allies, and our enemies has forever exploded the myth that what a people can produce is limited by funds in hand or in sight."[27] It disturbed him to think that a free society could muster resources for destruction, but not for peace.

He believed strongly that the nations of the world had to change their economic policies if they ever hoped to bring an end to war. In the 1940s, he participated actively in the National Planning Association, a private, nonprofit organization with standing committees on agriculture, business, labor, and national and international pol-

25. Thad Snow, "A Farmer Looks at Fiscal Policy" (Washington, D.C.: National Planning Association, Pamphlet No. 48, 1945), 5–6.
26. Ibid., 21, 22.
27. Ibid., 16.

icy. Under the auspices of this group, he produced his pamphlet entitled "A Farmer Looks at Fiscal Policy," published in 1945. Admitting his "extreme fiscal youthfulness,"[28] he urged other, more sophisticated people to take up the challenge of creating postwar economic stability.

Even in the throes of depression and war, he believed that human beings had the ability to solve their problems. These disastrous events proved economic, social, and political institutions lagged "far behind the facts of life—perhaps farther—than they have even lagged before in the history of the race." Modern economists needed to be wary of repeating the mistakes of their predecessors, but they should not react with "overcaution and inaction." If people applied intelligence and ingenuity to the social sciences, they could find the answers they needed. If not, disaster would surely come: "The trick we've got to learn is to amend our institutions before they fall down around us, to bring our ideas and habits up abreast with changed facts before they drive us to war abroad—and just as probably into bloody disorder at home."[29]

Snow believed that mankind's only hope for the future rested in a growing commitment to nonviolence. For him the touchstone was the sharecroppers' roadside strike of 1939. In his memoirs, he compared Whitfield's marchers to Gandhi's famous Salt March to the sea in 1930. As a public dramatization of the effects of oppression, the Indian salt marchers made camp in temperatures well above one hundred degrees. Nine years later, the Bootheel demonstrators slept in miserable camps in rain and snow.[30] Both of these peaceful protests drew worldwide attention to poverty and injustice, which Snow believed were the roots of war. The success of these demonstrations convinced him that nonviolence was not only possible, but also powerful.

After the roadside strike, he became increasingly convinced that social and economic justice formed the foundations for world peace. Writing as a farmer in 1945, he commented on the interdependence of all economic groups. Any special advantage for one group— whether farmers or businessmen—meant a disadvantage for other

28. Ibid., 1.
29. Ibid., 22.
30. Snow, *From Missouri*, 275–76.

groups.[31] War, he believed, resulted from ruling elites trying to gain or to keep their power and wealth at the expense of the underclass. Citing Veblen in another essay, he concluded that the common man benefited least and suffered most in a war.[32] Tariffs and protective trade policies, aimed at promoting one nation's economic health at the expense of other nations', were preliminary acts of war. If people and nations were ever to achieve lasting peace, they would have to live together on the basis of equality and fairness. He realized that this would require immense changes in human institutions.

Snow was not totally alone in his beliefs. Some prewar pacifists resisted the pressure to embrace militarism. For instance, A. J. Muste, leader of the Fellowship of Reconciliation (FOR), declared in 1942 that the world war would lead either to a breakdown of Western civilization or to a strong nonviolence movement. In order to build this movement, he began making contacts with oppressed minority groups, including African Americans. Muste and his associates, many of whom had strong ties to Protestant churches, envisioned the application of Gandhi's methods to fight injustice in America.[33]

As in years past, Snow's books gave him solace, and his family sheltered him. If he felt isolated, he could always draw upon Tolstoy, Veblen, and the writings of Gandhi for support. His daughters Fannie and Emily embraced his commitment to nonviolence. But others in the Bootheel were less sympathetic.

In January 1944, he returned home from Washington, D.C., to find his family alarmed at an unsigned note someone had shoved under his door. The writer warned him to change his position on the war or leave the Bootheel. "If you love the japs more than you do your own people," the note threatened, "you better get out while you can."[34]

Apparently, the writer subscribed to the *St. Louis Post-Dispatch*, because he or she demanded that Snow publish a recantation in that paper. In light of events yet to unfold, it seems possible that the threat came from a relative or an associate who was deeply embarrassed by Snow's public antiwar stance and wanted a retraction in print.

31. Snow, "Farmer Looks at Fiscal Policy," 5.
32. Snow, "Thorstein Veblen and the Nature of Peace," 4–5, Snow Papers, folder 57.
33. Wittner, *Rebels against War*, 12, 63–64.
34. Snow Papers, folder 62.

Priscilla's husband, Hartwell Thompson, had harbored hostile feelings toward his father-in-law since the early 1940s. Apparently, Snow contributed to those feelings by constantly haranguing and belittling Thompson.[35] The younger man suffered from bouts of depression that would ultimately have tragic consequences.

Typically and ill-advisedly, Snow responded offhandedly to the anonymous note. He made light of the threat and dismissed the writer's demands as ludicrous: "I think I like the folks around me fully as well [as the Japanese]. But it does seem right silly to say it, and unscientific, too, because I don't happen to know any Japanese."[36]

In a time of national crisis, when nearly all of his fellow citizens supported America's war effort and many died in that effort, he should have concerned himself more with the difficult position in which he had placed his family. It could not have been easy to have a father or a father-in-law who seemed to be more sympathetic with the nation's enemies than with its fighting men. The tensions of the 1939 roadside strike had only partially abated when Snow took his wildly unpopular and seemingly unpatriotic stand in 1941. Although he could make light of it, perhaps those closest to him could not.

In truth, he took the war and its consequences very seriously. He cared about his neighbors in the Bootheel, and he was not disloyal to his country. But he feared that violence and militarism might bring the end of the human race. "Why should I be concerned about the future of my species? I would say it is because we are an interesting species."[37] By the late 1940s, he knew that human beings possessed the ingenuity to create horrendous weapons of destruction. No one could predict what the atom bomb might do to the world. He hoped the new weaponry might at least make people stop and think. In the past, men had settled disputes by fighting duels. When firearms became more accurate and deadly, the custom faded away. Perhaps the old habit of war would meet with the same fate.

35. Wallhausen interview, October 28, 2002.
36. Snow Papers, folder 62.
37. Snow, *From Missouri*, 341.

Ozarks Retreat 9

Here's hoping you come back to Van Buren this summer so that you and Rip and Ben and I can sit around a campfire and settle the fate of the world.

—Leonard Hall, *Charleston Enterprise-Courier,* January 6, 1955

Near the end of his life, Snow left the Bootheel and lived in the Ozarks. According to George Burrows, he kept a bedroom at Snow's Corner but spent most of his time at a hotel in Van Buren.[1] He did this for many reasons. After struggling with the manuscript for more than ten years, he needed a quiet place to work on his book. The Depression, the war, and his intense preoccupation with economic and political policy had drained him. A quiet life in the mountains offered a respite, if not an escape, from the "war-crazed world."[2] But there was another, more personal, reason for his retreat from the Bootheel.

On Tuesday, August 10, 1948, Snow and his daughter Emily returned from an extended trip to Australia and New Zealand.[3] After this trip, Frank O'Hare said, "Thad was in fine form, in far better spirits than when he started on his seven months tour of the world."[4] Emily, a student at the University of Mexico, was her father's frequent companion. She had a degree from the University of Wiscon-

1. George Burrows, interview with author, Van Buren, Mo., July 10, 1998.
2. Snow, *From Missouri,* 40.
3. Warren interview, November 7, 2002.
4. Frank P. O'Hare to John [Stewart], August 19, 1948, O'Hare Papers.

138

sin and was enrolled as a graduate student at Radcliffe College, Cambridge, Massachusetts, for the fall term. Friends viewed the father and daughter as inseparable "pals."[5]

At this time, Emily was the most important woman in Snow's life. According to Millie Wallhausen, Snow had little communication with his son, Hal, and seemed to pay little attention to his daughters Priscilla and Fannie. By this time his two older daughters had married and turned their attention to their own families. "Emily was the one he adored," said Wallhausen. "He took her everywhere with him."[6]

Emily's best friend, Julia Cooper Warren, described the relationship between Snow and his children in similar terms. He and his son, Hal, had little in common, she said. Priscilla was dark-haired, pretty, proper, and very reserved. Fannie was "the wild one," independent, an expert horsewoman. Emily, blonde and pretty, "was her father's favorite," Warren said. "But that went two ways. He was *her* favorite. And what she meant to do was to please him. Everything she did was to please her dad."[7]

While Emily was growing up, she and Julia spent a lot of time at Snow's Corner. They went to grade school and high school together and would often spend nights and weekends on the farm. Warren recalled that Snow always had time for them. He let them take books from his library, and he would tell them what books to read. "Emily was very earnest about books," she said. "I would argue [with Snow]. I was a real smart aleck. But he liked that."[8]

According to Warren, Emily rarely questioned her father. What he believed, she believed.[9] She shared his dedication to social reform, and he hoped she would become an activist. Frank O'Hare expressed this in verse:

> She was [his] gift to you, World.
> Dedicated. Preparing herself to serve You.
> Not with ambition, but simply. Think of the
> Women whose names are writ large in the hearts of men,

5. *Sikeston (Mo.) Standard,* August 16, 1948; letter to Frank P. O'Hare, signed Elizabeth, August 24, 1948, O'Hare Papers.

6. Wallhausen interview, October 28, 2002.

7. Warren interview, November 7, 2002.

8. Ibid.

9. Ibid.

Whose love was like the Christ's, who lived in the
World that is to be, builders of that world,
Increasers of freedoms and joys, destroyers of false gods,
Along whose pathway roses always grow.
So was Emily to be.[10]

But Emily's life ended abruptly and brutally, before she had a chance to realize her potential.

On Friday evening, August 13, her friend Julia visited Snow's Corner, staying until about eight o'clock.[11] Priscilla's husband brought the newspaper to the house at about ten. He and Snow had a friendly conversation for about half an hour, and then Thompson went back to his own house just across the road.[12]

Early the following morning, Thompson would kill his wife, daughter, and sister-in-law, and then shoot himself in the head with a .22-caliber rifle. His forty-year-old wife, his nine-year-old daughter, Ann, and her Aunt Emily would die on the Thompson property, just a short distance from the farmhouse where Snow was sleeping.

Before shooting himself, Thompson phoned the county coroner, John F. Nunnelee Jr.[13] The approximate time of the call was 6:55 on Saturday morning. Coroner Nunnelee said later that Thompson apparently killed his wife and daughter, who were sleeping together, and then killed Emily. He did not turn the gun on himself until after he notified authorities. The coroner discovered him at the foot of his wife's bed, still breathing, but fatally wounded. Thompson would die a few hours later in St. Mary's Hospital in Cairo, Illinois.

Possibly Emily heard gunshots and went to the Thompson residence to see what was wrong. Some neighbors believed her brother-in-law actually phoned or went and awakened her, luring her to her death.[14] Her friend Julia thought that Hartwell must have come to her and told her that Priscilla, who suffered from frequent headaches, needed her help.[15]

---

10. Frank P. O'Hare, "Emily: A Father's Lament (For Thad Snow)," O'Hare Papers.

11. Warren interview, November 7, 2002.

12. *St. Louis Post-Dispatch,* August 15, 1948.

13. *Cape Girardeau Southeast Missourian,* August 14, 1948.

14. Stallings interview, April 1, 2001.

15. Warren interview, November 7, 2002.

The front door of the Thompson house was locked. When Emily approached the rear of the house, her brother-in-law fired a shot through the screen door, leaving a bullet hole. The coroner concluded that he shot her while he stood inside as she walked toward the steps of the back porch. With wounds in the heart and leg, she appeared to have staggered backwards ten or fifteen feet before falling.[16]

After the shooting, Thompson crossed the road, heading for the Snow residence. James Manker, Snow's right-hand man, saw Thompson turn around and hurry back to his own house. Manker heard no shots and reported that Thompson was not carrying a gun. Possibly, he intended to kill Snow, but apparently, he went back into his own house and called the coroner.

Nunnelee vividly recalled the conversation.

Thompson asked, "Who is this?"

Nunnelee identified himself.

Thompson gave his name and then said, "I've just killed my wife, my daughter, and Emily Snow," then paused, caught his breath, and added, "Come and get us."

Nunnelee was five miles away in Charleston. He shouted into the phone, "Hartwell!" at the top of his voice. But it was too late to prevent the tragedy.

At the time of the shooting, Snow was in bed. Sheriff Walter Beck and a neighbor, Ruel Swank, awakened him.

Snow said, "My God, why didn't he kill me instead?"

Thompson, aged fifty, was a native of Charleston and a member of a prominent local family. He worked as a civil engineer in Dexter (Stoddard County) and in Paducah, Kentucky, before settling on the farm in Mississippi County. After his health failed in the early 1940s, he operated a 120-acre farm and a 280-farm adjoining the Snow property. According to Julia Warren, Snow helped him financially.[17]

Friends reported that Thompson suffered from depression and had previously talked about suicide. He had been under observation at the state mental health facility in Farmington, but had never been admitted. Apparently he had told acquaintances that he intended to kill his wife and daughter.[18] Julia Warren said of Priscilla,

16. *Sikeston (Mo.) Standard,* August 16, 1948. Much of the following information is from this article.

17. Warren interview, November 7, 2002.

18. *Cape Girardeau Southeast Missourian,* August 14 and 16, 1948; *St. Louis Post-Dispatch,* August 15, 1948.

"She only thought if he did something it would only be to himself. Never to this child that they had wanted for so many years to have. And they just loved this little girl. She was a darling little girl, too." Of Thompson, she said, "He was mentally ill. He was a Thompson. He came from a very prominent family. And that was in the days when if you were mentally sick you did not tell anybody. . . . He always had something against Thad. I never knew why. I never heard him [Thad] say anything unkind about Hartwell."[19]

Millie Wallhausen said Thompson was "an ordinary guy, from a good family, but not quick on the draw." She believed that he suffered because he could not match wits with his father-in-law. In her opinion, Snow badgered and belittled Thompson. Priscilla was a shy person, not the fighting type, and she could not defend her husband. Eventually, Thompson just snapped.[20]

Sheriff Beck, who accompanied the coroner to the scene, told reporters that the shooting resulted from many years of ill-feeling between Thompson and his father-in-law. Authorities found a suicide note, scrawled in black crayon, that said, "I loved my girls [space] To face it all alone and knew I was losing my mind—Now they are at rest.

"Thad Snow you will get some of the torture you have caused us." The word "me" was written above the word "us." The note continued, "You could have prevented this. God forgive."[21]

Neighbors remembered the pain it caused the family. Aunt Alma Catt, who had helped to rear Priscilla, was the cook at the rural school in Concord. Sometimes she brought her grand-niece, Ann, a frail little girl, to school with her. When Ann died, Catt was heartbroken.[22]

More than half a century later, Julia Warren said, "I've never recovered from Emily's death. I never have, and I never will."[23] Millie Wallhausen said, "I can still hear that phone ringing. We were at our dining-room table, when we were told what happened. I was speechless. It was Hartwell's revenge. [By killing Emily], he took the one thing Thad would have given his life for. She was a brilliant girl. We

19. Warren interview, November 7, 2002.
20. Wallhausen interview, October 28, 2002.
21. *Sikeston (Mo.) Standard,* August 16, 1948.
22. Stallings interview, April 1, 2001.
23. Warren interview, November 7, 2002.

didn't see it coming. Hartwell just couldn't compete with the mind of Thad."[24]

Frank O'Hare wrote an eloquent eulogy in a lyric poem entitled "Emily: A Father's Lament (For Thad Snow)." He identified with Snow's grief and expressed it in these lines:

Dear World—

Emily is gone.
Snatched away from me, from you
At the very threshold of her life.

How she responded to your cry, enslaved men,
And rejoiced in your valor, heroes.
She loved Truth and Justice.

Not to every father is it given to have a daughter like Emily.
Nor, like me, to have her struck down.
What shall I do in my agony?[25]

Emily's death brought her siblings home. Snow's son, Hal, came from New York to attend the funeral services on Monday, August 16. Snow's remaining daughter, Fannie, and her husband, Bob Delaney, traveled from Georgia with their three young children and stayed to run the farm. The Delaneys never left the Bootheel again.

Life at Snow's Corner was changing. In the late 1940s, there were still black sharecroppers on the farm. Snow had lobbied in Washington to reform the sharecropping system, not to end it. He did not dismiss his own croppers or ask them to work as wage laborers. According to Fannie Delaney, they were allowed to "retire" on the land, but "they drifted away, to Charleston mostly," as time went by.[26]

Alex Cooper, the son of an African American sharecropper, believed that the end of the sharecropping system was a mixed blessing. Day laborers and wage workers had no roots in the land; they merely sold their skills to the highest bidder. In the 1950s and 1960s, mechanization forced the croppers to become wage workers, and

24. Wallhausen interview, October 28, 2002.
25. O'Hare, "Emily."
26. Delaney interview, March 19, 2001.

they moved to the cities for economic reasons, not necessarily because they wanted to go. The result, according to Cooper, was an erosion of a strong sense of family and a cohesive rural black community.[27]

For a few years in the late 1940s, the Delaneys continued farming with sharecroppers and mules, but gradually they turned to more modern methods. In the 1950s, they acquired tractors and other sophisticated machinery. Eventually, Fannie's daughter Debbie and her husband, Wayne Corse, took over the farm and did most of the work themselves with the help of modern equipment.[28]

With the younger generation in charge, Snow escaped to Van Buren to write his book. Fannie said that her father found it impossible to concentrate at home while she and her young family were living at Snow's Corner. George Burrows, who knew Snow in the 1950s, said he went to the hill country to escape from the tragedy ("you know, the tragedy"), although he never talked about it.[29] According to Fannie, he went because he truly loved the Ozarks. He had probably become familiar with the hills on hunting trips or while visiting Whitfield at Cropperville.

After forty years of clearing and farming flat land, he finally went back to the woods. In the early twentieth century, loggers had harvested most of the timber in the Ozarks, but the cleared land was rocky and unprofitable for farmers. During the Depression, the Civilian Conservation Corps planted hundreds of thousands of trees in a massive reforestation program. The United States Forest Service and the National Park Service acquired huge tracts of land, protecting them from private development.

Van Buren was nestled in the rugged Courtois Hills, about thirty miles north of the Arkansas line and about sixty miles west of Poplar Bluff. With a population of about nine hundred, it was the largest town and the governmental center of Carter County (which held a population of five thousand).[30] The rubble stone courthouse dominated a square, surrounded by banks, drugstores, service stations, tourist shops, and restaurants.

27. Alex Cooper interviews, March 11 and July 27, 1994.
28. Corse interview, May 28, 2002.
29. Burrows interview, December 15, 1998; Delaney interview, August 16, 1999.
30. Jimmy Lowe, *Striving Upward: An Autobiography*, 136.

Because of its location at the break line between the upper and lower sections of the Current River, Van Buren was a popular starting point for guided float fishing trips downstream. In the 1870s, John Emerson, of Ironton, Missouri, founder of Emerson Electric Company in St. Louis, came to the area and became its chief promoter. He stocked the Current with fish from federal hatcheries. Partly due to his efforts, Van Buren became a well-known sportsmen's retreat.[31]

Snow took rooms in the Rose Cliff Hotel, a clean, commodious, but rustic inn on a promontory overlooking the river. Dr. Robert Davis of Shannon County constructed the hotel in the late 1920s, during the lumber boom in the Ozarks. He located the facility on the south bank of the river very close to the new steel bridge on Highway 60. A footbridge near the highway bridge allowed guests to cross the river into the town. Tourists were attracted to the hotel because the state had recently opened three new state parks, Big Spring, Round Spring, and Alley Spring, in the valleys of the Current and Jack's Fork Rivers.[32] At times, the place was so crowded that guests slept several to a room, in the lobby, or on the verandas in sleeping bags.

From the 1930s through the 1950s, the Rose Cliff was a magnet for salesmen, sportsmen, recreational tourists, writers, and conservationists. Hotel manager Ben Davis was an amateur naturalist. During the Depression, federal officials and surveyors who laid out Clark National Forest stayed at the hotel. In the late 1930s, members of the newly created Missouri Conservation Commission met there to plan for preservation of the state's wildlife. Environmentalists met there to oppose plans to dam the Current River and develop counterproposals to create what would eventually become the Ozark National Scenic Riverways. Writer, farmer, and naturalist Leonard Hall stayed there in the 1950s, and for many years, it was a hangout for *Post-Dispatch* staffers.[33]

Snow had a basement room with windows looking out on the river. His friend Leonard Hall described the river's beauty:

31. Lynn Morrow, "Rose Cliff Hotel: A Missouri Forum for Environmental Policy," 39–41.
32. Ibid., 40–41.
33. Ibid., 41–43, 47.

> The Current River, which flows through the eastern Ozarks, is the loveliest of all our Missouri streams and can hold up its head for beauty alongside any on our North American continent. I like best its old French name, *La Riviere Courante*—which is to say, "Running River"—for nothing could better describe the swift course it takes down its wild and narrow valley.[34]

Snow did not remain shut away in his room, but wandered through this valley and the surrounding hills.

Ben Davis, his host at the Rose Cliff, drove hundreds of miles with him on narrow roads through Carter County. Most of the land belonged to the Forest Service. There were few houses, but many old clearings, where people had tried to establish farms and then moved on.[35] Agriculturally the Ozarks remained marginal, with row crops limited to flood-prone creek bottoms and rangy livestock roaming hills and plateaus. Modern mechanized farming did not flourish in the region, and the farmers who stayed there often lived in poverty.[36]

From Davis, Snow learned the geography and lore of the hills. "The hills are mostly without names," he wrote, "possibly because there are so many of them and not enough names to go round. But the valleys and hollows all have names, or at any rate all those that have ever been cleared for the plow." Davis told him the story of Horse Hollow, which was so narrow that a poor horse trying to jump a fence on one of the ridges kept missing and ending up on the opposite ridge. Trying to get to a luscious row of corn, he kept jumping back and forth from one ridge to another, until he finally died of exhaustion. This was a fable that expressed the humorous affection the hillmen had for their land and its rugged terrain.[37]

Snow loved the names of places in the Ozarks. In his memoirs, he told the story of the post office at Rat, Missouri. Jake Lister, the postmaster, wanted to name it "Alice," after his youngest daughter. Not wanting to play favorites, however, he first submitted the name of his eldest daughter, Edith. After several arduous trips to the post of-

---

34. Leonard Hall, *Stars Upstream: Life along an Ozark River*, 1.

35. Snow, *From Missouri*, 44–45.

36. Rex R. Campbell, John C. Spencer, and Ravindra G. Amonker, "The Reported and Unreported Missouri Ozarks: Adaptive Strategies of the People Left Behind," 31–33.

37. Snow, *From Missouri*, 45–46.

fice in the next town, he received a negative reply from the United States government. There was already a Missouri post office named "Edith." So he submitted the name of his second daughter, Ethel. Again the reply was negative, so (secretly rejoicing) he submitted Alice's name.

In the middle of winter, Lister trudged to town, walking for two days in sleet and snow, only to receive a third negative reply. Outraged and discouraged, he wrote in big letters the word *RATS* on the official form. In the spring, he received a positive reply. Thus, the new post office was named "Rat."[38] This story must have resonated with Snow, who was, after all, a man who had had three daughters and a special bond with the youngest.

He had a genuine respect for the people of the Ozarks. Others might have looked down on them for their lack of material wealth and sophistication, but Snow wrote, "I suspect the average I.Q. is as high as in some more populous and prosperous areas, where the business of living is more complicated, and where you can't drive out four miles from town as I did last evening and watch twelve deer browse unafraid in a little glade."[39] He believed his Ozark neighbors had made a conscious, rational choice to preserve their connection with nature rather than pursue success, defined in terms of money and possessions.

His second sojourn in the wilderness gave him a chance to look back on his life in the Bootheel. In the early days at Snow's Corner, he had lived on a wooded ridge in a swamp, which was a haven for wild turkey, panthers, and snakes. Late in his life, he remembered putting on a show for his children. Riding across an open field on his horse, he swung down and picked up a six-foot-long blue racer and twined it around his saddle horn. When he got home, his children and some neighbor children were playing in the yard. He let the snake twine itself around his arms and neck, with its mouth only a few inches from his face. Later, he let the children play with the snake. In the evening, he took it back out into the field and set it free.[40]

38. Ibid., 43–44.
39. Ibid., 41. It is important to note that in the 1950s, deer populations were very low in Missouri due to lumbering, overhunting, and poor wildlife management.
40. Ibid., 145–50. Julia Warren confirmed that the Snow children were not afraid of snakes, although she was (interview, November 7, 2002).

Observing wild creatures caused him to think about human nature. He believed that people's aversion to snakes was a learned response, not an instinctive one. For him, there was a lesson in it. He
wrote that, "Snake hatred is a shameful thing, like race prejudice
and race hatred. We are not born with these regrettable notions, but
acquire them along with all the other good and bad notions that
make up our culture."[41] Through several decades of struggle in the
Bootheel, he became painfully aware of the flaws in American culture.

From his vantage point in the mountains, he watched the world
tumble into the Cold War. After Roosevelt's death in 1945, Harry S.
Truman of Independence, Missouri, guided the United States through
the last days of World War II and attempted to define peacetime foreign and domestic agendas. In foreign affairs, Truman relied heavily
on advisors like James Byrnes, George Marshall, and Dean Acheson, who shared a strong antipathy toward the Soviet Union.[42] On
the domestic side, the president faced sharp opposition from con-
servative Republicans, who wanted to reduce the size of the federal
government and dismantle the New Deal.

In 1947 and 1948, the House Un-American Activities Committee
(HUAC) investigated the entertainment industry, federal agencies,
Roosevelt appointees, and a host of political activists in an attempt
to expose possible communist influence on the New Deal. The General Records of the Department of Justice, Record Group 60 in the
National Archives, contain a reference dated March 13, 1947, to national security matters relating to Thad Snow as an alleged member
of the Communist party. As previously stated, according to the Department of Justice, this file was destroyed in 1991. While HUAC
validated its investigations by invoking phrases like "national security," its true agenda was domestic. Committee members sought to
discredit the ideals and principles of the Roosevelt years.

In August 1948 Whittaker Chambers, a senior editor of *Time* and a
former communist, swore to HUAC that Alger Hiss, a State Department official, had been a member of a communist organization.
Chambers also spoke out on radio, and Hiss sued him for slander.
In November and December of that presidential election year, in

41. Snow, *From Missouri*, 150.
42. Athan Theoharis, *Seeds of Repression: Harry S. Truman and the Origins of
McCarthyism*, 34.

which Truman defeated Republican Thomas E. Dewey, Chambers produced additional documents linking Hiss to communist espionage. Representative Richard M. Nixon, a member of the HUAC, signed a subpoena ordering Chambers to turn over all relevant materials to the committee. In a bizarre scene, Chambers led investigators to a Maryland pumpkin patch, where he pulled three canisters of incriminating film from a hollowed-out pumpkin.[43] Truman publicly expressed contempt for HUAC's investigations, but the Hiss affair cast a cloud over his administration and the State Department.

Ironically, and fatefully, Truman's own anticommunist rhetoric increased the level of suspicion. In 1949, when Truman sought ratification of the North Atlantic Treaty Organization (NATO), he identified the Soviet Union as the primary threat to peace in the postwar world. America, in his view, was the principal protector of western democracies. Soviet agents, he alleged, were operating throughout the world in an attempt to create disunity and to undermine America's military defense program. Further, he stated, communism denied the existence of God, threatening both liberty and faith. In essence, he portrayed a completely implacable and evil enemy, with whom compromise would be impossible. Without a strong military alliance of western powers, war would be inevitable.[44]

During this perilous time, Snow sought tranquility, but he could not cut himself off entirely from the wider world. In the late winter and early spring of 1950, he spent several weeks at Harvard University, attending seminars. Returning to Van Buren from that trip, he tried to regain his composure. "The hills are alluring and reassuring," he wrote, "but they calm you down only gradually."[45]

He believed that the world had gone mad. In particular, he believed that Truman and his new secretary of state had succumbed to a kind of paranoia or "diplomatic dementia." In the spring of 1950, Acheson issued a top-secret document, NSC 68. This document warned that U.S. military strength was becoming dangerously inadequate, since the Soviet Union now possessed the atomic bomb. Urging a rapid buildup of military strength, NSC 68 argued that the

43. Robert J. Donovan, *Tumultuous Years: The Presidency of Harry S. Truman, 1949–1953,* 32–33.
44. Theoharis, *Seeds of Repression,* 55.
45. Snow, *From Missouri,* 40.

Cold War was in fact a real war, and the survival of the free world was at stake.[46] Publicly, the secretary of state set the stage for a millennial conflict between the western democracies and implacable communist dictatorships bent on destroying Americans' minds and spirits. To fight this threat, he advocated "total diplomacy," a nebulous but ominous concept, implying an absolute commitment to the destruction of communism. This, Snow was convinced, would inevitably lead to war.[47]

Acheson was an elegant man with prominent eyebrows, intense brown eyes, and a dramatic mustache. In contrast to the Missouri-born Truman, the secretary of state was a New Englander, educated at Yale College and Harvard Law School. During Roosevelt's tenure, he had served as under secretary of the treasury and later as assistant secretary of state, returning to his law practice in 1947. He joined Truman's cabinet in 1949. According to Snow, he was "a conscientious, zealous public servant, handsome in appearance, well dressed and insane."[48]

From the point of view of "an American citizen who lives deep in the Ozark hills," Snow found Acheson's speeches incendiary and incomprehensible. The words "total diplomacy" had no meaning, but Snow did get the idea that Acheson was asking the American public to "approve in advance any and every adventure of the State Department." This, he implied, was a necessary and not excessive response to a subtle but terrible menace that threatened not only our country but also "our self-respect and the integrity of our mind and spirit." Acheson used frightening words, meant to instill fear and arouse people's passions. And, according to Snow, they "made us all properly pop-eyed, except for the hill people I talk with, and who don't count."[49]

Personally, Snow said he was scared under the bed. But he came out of hiding when he considered the implications of Acheson's philosophy. Communism posed no threat to Snow's self-respect or integrity. These things were his own. In his words, the secretary of state could not "scare me about these things, because he can do absolutely nothing about them, one way or the other. Nor can Mr. Joseph Stalin."[50]

46. Donovan, *Tumultuous Years*, 34–35; Snow, *From Missouri*, 48.
47. Snow, *From Missouri*, 46–49.
48. Donovan, *Tumultuous Years*, 34–35; Snow, *From Missouri*, 48.
49. Snow, *From Missouri*, 47–49.
50. Ibid.

Snow correctly predicted that Truman and Acheson's overblown warnings about the communist menace would arouse emotions that would be hard to control. In late February 1950, Joseph McCarthy, a Wisconsin Republican, held the floor of the U.S. Senate for six hours, while he accused eighty-one anonymous employees of the State Department of being communists.[51] Twisting the president's own rhetoric, McCarthy embellished the idea that there was a worldwide conspiracy of communists who were acting secretly and insidiously to undermine the American character. He further argued that vast numbers of communists had infiltrated the U.S. government and that the president himself was a tool of these "twisted intellectuals who tell him what they want him to know."[52] Stunned by McCarthy's harangue, Democratic leaders ordered an investigation by the Senate Foreign Relations Committee, opening the door for continuous and escalating rhetoric by the man from Wisconsin.

In an atmosphere of fear resulting from the threat of atomic warfare, the fall of China, military action in Korea, and media-induced suspicion of subversion, McCarthy attracted a huge national following. The Hearst, Scripps-Howard, and McCormick newspaper chains endorsed his crusade. Columnists and radio personalities quoted him. Contributions flowed into his office from manual laborers, farmers, small businessmen, the American Legion, the Veterans of Foreign Wars, Republicans, Democrats, fringe groups, and hate groups. McCarthy obliged them with continuous tirades, leveling reckless accusations, and inspiring suspicion within the government, private industry, the arts, and the entertainment business.

McCarthyism did not spare the Bootheel. In September 1952, FBI agents arrested Marcus Alphonse ("Al") Murphy, an African American laborer and former member of the Communist party, at his home in Charleston. The U.S. government charged him with violating the Smith Act, a 1940 law making it a crime to join an organization that taught, advocated, or encouraged overthrowing the government by force. Murphy's alleged crime was teaching classes in communism.[53] The Smith Act resulted from American fears of totalitarianism in the years before World War II. During the McCarthy

---

51. Donovan, *Tumultuous Years*, 164.
52. Ibid.
53. Joan Tinsley Feezor, "Marcus Alphonse Murphy and Communism on Trial: The Smith Act in Missouri," 1–7.

era, FBI agents and federal prosecutors invoked the Smith Act in order to round up radicals and so-called radicals from coast to coast.

Murphy had played a minor role in the sharecroppers' strike in 1939. Born in Georgia in 1909, he grew up in abject poverty and finally found employment as an industrial worker in Birmingham, Alabama. In 1930, he joined the Communist party in that state, where he helped organize the SCU. As a member of the party, he traveled to many places, including the Soviet Union.[54] During the Depression, he worked as a party organizer, lecturer, and teacher in St. Louis. In all likelihood, he was the "Comrade M" who taught classes at the communist school for sharecroppers in St. Louis in 1934.[55] When he heard about the roadside strike, he volunteered to go to southeast Missouri. There he met his future wife, Pauline Hawkins Butler of Charleston. Using his party connections, he helped collect money, food, and clothing for the Committee for the Rehabilitation of the Sharecroppers.

Returning to St. Louis, Murphy remained active in the Communist party. An energetic, articulate man, he took part in struggles against segregation, unfair labor practices, and police brutality. In 1940, he ran for lieutenant governor on the Communist party ticket. During World War II, he was active in civic work in St. Louis.[56] A decade later, he protested against U.S. involvement in the Korean war. By 1952, however, he left the party and moved to Charleston.

There he settled down with his wife in a white frame home on a county road. Pauline was a teacher in rural black schools. Murphy held various odd jobs and reportedly was intending to start work at a Cairo car dealership when FBI agents arrested him at his home early on the morning of September 17. On that same day, agents also apprehended seventeen other alleged communists, including William Sentner, a former party leader in St. Louis. Convictions under the Smith Act could result in prison terms of up to ten years and fines of up to ten thousand dollars.[57]

Two years later, Murphy and four other defendants from the St. Louis area endured prosecution in court for violating the Smith Act

54. Ibid., 27, 54–55, 71–74.
55. Reel 287, folder 3714, CPUSA Records.
56. *St. Louis Argus*, September 19, 1952.
57. *Charleston (Mo.) Enterprise-Courier*, September 18, 1952.

and for agitating to overthrow the government. During the five-month trial, prosecutors provided no evidence that Murphy had ever participated in or advocated violent revolution. Nevertheless, the jury convicted him and all the other defendants, who were released on bail pending an appeal. In 1957, after the McCarthy hysteria subsided, all five defendants were granted a new trial, and in the fall of 1958, the charges were dropped.[58]

Thad Snow contributed ten thousand dollars for Murphy's bail.[59] As the case dragged on, Snow spent much of his time in Van Buren, writing his book. By the time Murphy won his freedom, Snow had died. In 1958, when the case finally ended, L. A. Simpson, the executor of Snow's will, collected the ten thousand dollars and distributed it to his heirs, Hal Snow and Fannie Delaney.[60]

Although he had retreated to the Ozarks, Snow could not turn his back on injustice. In a sense, near the end of his life, he sought out a new frontier, but the frontier had a different meaning for him. It was no longer a place for a young man to flex his muscles, grapple with nature, and create a new world by the strength of his will. Over the course of many years, he learned that the society he helped establish in the Bootheel was not the new creation of courageous men, but the sad reflection of old, outmoded systems, steeped in racism and injustice, nurtured in the ancient institution of slavery. When he came to the Ozarks, he was prepared to accept a new frontier on its own terms, appreciate its wildness, and absorb its quiet lessons.

He was afraid that, for many Americans, these lessons no longer had meaning. Like the old man in John Steinbeck's "The Leader of the People," he believed that Americans no longer understood the pioneering urge. The frontier had vanished from the United States. But, more significantly, it seemed to have disappeared from peo-

58. Feezor, "Marcus Alphonse Murphy," 159.
59. Ibid. The *St. Louis Globe-Democrat,* January 1, 1953, confirms that Murphy's bail was set at ten thousand dollars.
60. Although Thad Snow's estate was settled on May 18, 1956, in 1958, L. A. Simpson, executor of the will, collected additional assets in the amount of $10,000. Because of the recovery of these additional assets, the estate was reopened. After the taxes were paid on these new assets, Hal Snow and Frances Delaney each received $3,151.88. Snow's probate file is missing from the Mississippi County Courthouse. However, the Order Approving Final Settlement and Order of Distribution, dated July 24, 1959, were included in the abstract of Snow's property on file at Mississippi County Abstract and Loan in Charleston, Missouri.

ple's consciousness, as Steinbeck's old man says to his grandson, Jody: "No place to go, Jody. Every place is taken. But that's not the worst—no, not the worst. Westering has died out of the people. Westering isn't a hunger any more. It's all done."[61]

In the 1950s, Snow saw a world that had given in to bigotry and fear. Although he looked for solace in the Ozarks, he did not withdraw from the world. Out-of-town visitors came to his rooms at the Rose Cliff for afternoon toddies and long conversations. Old friends like Daniel Fitzpatrick, political cartoonist for the *St. Louis Post-Dispatch*, kept him company and shared the news of the day. As always, he read eclectically from popular magazines, academic tracts, Gandhi, and Veblen.

He had paid a price for expressing unpopular ideas in a world that had no tolerance for unconventional thought. He came to the aid of Al Murphy when the world tried to punish him for speaking and teaching what he believed. The real frontier, Snow knew, was not at the edge of the settled landscape, but at the edge of acceptable and accepted thought. The real frontier was in the mind.

61. John Steinbeck, "The Leader of the People," in *The Long Valley*, 225. Julia Warren confirmed that Snow was a reader of Steinbeck (interview, November 7, 2002).

From Missouri 10

In much of what he believes Thad Snow is "from Missouri." In putting that on paper he has written a vivid piece of Americana, as indigenous to the black soil that produced it as the mulberries that shade the land and drop their fruit on it.

—W. C. White, *New York Herald Tribune,* November 21, 1954

In his last years, Snow sought peace, but he did not become a recluse. Many visitors came to his farm in Charleston or to his room at the Rose Cliff Hotel, where he listened to their opinions, expressed his own views, and told humorous stories. Because of his weak legs, he liked to talk, study, or read lying on his bed.[1] Suffering from insomnia, he did most of his reading and writing at night.

He wrote on any available surface, including scraps of old letters, legal pads, business forms, and the backs of envelopes. Although he sometimes prepared a typed copy of his compositions, he insisted that he could only retain the flow of his thoughts by writing in longhand.

His book, *From Missouri,* was published in the fall of 1954. On November 8, he autographed copies of the book at a St. Louis department store. Owen Whitfield and his daughter Barbara were there. During the event, Snow remained seated, and he walked with great difficulty. Because of his physical weakness, he returned

1. Julia Warren said, "Did you ever see anyone write while lying in bed? Well, that's what he did" (Warren interview, November 7, 2002).

to Charleston, but he was planning to go back to Van Buren and write a second book.[2]

Health problems, which began in the late 1930s with the paralysis in his legs, continued to plague him. For the last two decades of his life, he walked with a limp. In his sixties and seventies, he suffered a series of strokes. His condition had been worsening steadily for five months when he was admitted to St. Mary's Hospital in Cairo, Illinois, on January 4, 1955. He died there of pneumonia on January 15 at the age of seventy-three.[3]

His most enduring legacy was the odd, rambling memoir of his life in Missouri. Reviewers hardly knew how to categorize it. Helen Cain of the *Chicago Sunday Tribune* called it a "combination autobiography–social study." The *New York Herald Tribune Book Review* called it "a vivid piece of Americana." Victor P. Hass, writing for the *New York Times*, called Snow a "militant individualist" in "an age of conformity." The book, Hass commented, revealed Snow's intellectual sophistication as well as his "uncommon common sense," capacity for moral indignation, and sense of humor.[4]

One of his most sympathetic readers was the Ozarks environmentalist Leonard Hall, who wrote in a letter addressed to Snow and published in the *Charleston Enterprise-Courier* that he had lost three nights' sleep while he read the book from cover to cover. A farmer, Hall could read only in the evenings, after feeding and caring for livestock and finishing other chores. He could identify with Snow's experiences of clearing the swamps and surviving through floods, droughts, and depressions. His review of the book was personal, not professional. "It's a swell story—Thad—and I'm glad that

2. *Poplar Bluff (Mo.) Daily American Republic,* January 17, 1955; Fleming interview, September 10, 2002.

3. *St. Louis Post-Dispatch,* January 16, 1955. His will was recorded in Mississippi County deed book 164, p. 302, on January 26, 1955. Under the terms of this will, he left all his real estate to his daughter, Fannie Snow Delaney, with the remainder of his estate to be divided equally between Fannie Delaney and Hal Snow, who was living in Sedgwick County, Kansas. Snow's probate file is missing from the Mississippi County Court records. However, probate documents were copied and included in abstracts of Snow's real property. His estate was admitted to probate January 15, 1955, and settled May 18, 1956. The case was reopened in 1958 due to the discovery of the additional $10,000 in assets, discussed in Chapter 9.

4. *Chicago Sunday Tribune,* November 8, 1954; *New York Herald Tribune,* November 21, 1954; *New York Times,* November 14, 1954.

with all its vicissitudes, you had the courage and humor to end it on a high note."[5] Perhaps the most remarkable quality of the book, in light of the events it recorded and the recent tragedies in the writer's life, was its undaunted optimism.

Just what kind of book is it? In part, it is an autobiography. Although Snow did not present a coherent, complete, chronologically organized account of his life, he recalled events from his life as they related to his conversion to a radical view of politics and society in the middle twentieth century. Some of his recollections were self-congratulatory, as when he praised his own skills as a farmer. Others were self-critical, as when he discussed his "big-eye"—his desire for success, his ambition and greed. Others were probably self-deceptive, as when he said he became a "class-conscious farmer" as early as 1907.[6]

In part, it was a political monologue, a long letter to an imaginary editor, explaining his view of the world. He tackled the issues of war and peace, freedom of expression and political persecution, racial and class tensions, and economic justice. The topic to which he returned again and again—the embodiment of all his political views—was the Sharecroppers' Roadside Strike of 1939.

The book is also partially a collection of down-home stories about the colorful people of southern Missouri. One of the most revealing of these was the story of Budge Cobb, who single-handedly challenged the power of the U.S. Army Corps of Engineers. Cobb, a backwoodsman, came down to the levee during the great flood of 1937. The Corps had already decided to blast open the levee and flood out the farmers in the spillway. Cobb was not a farmer, and he cared very little about the army. He came down to the levee with his shotgun, hunting for rabbits. But the uniformed crew of soldiers misinterpreted his intentions.

Budge Cobb intimidated the U.S. Army. After carefully assessing his "startling whiskers and grizzly brows,"[7] the dynamiting party quietly withdrew from the scene and did not return until the following day. In this way, the backwoodsman beat back the Corps and delayed the breaching of the levee, giving the farmers and sharecroppers an extra day to flee to dry land. For Snow, the story of

5. *Charleston (Mo.) Enterprise-Courier,* January 6, 1955.
6. Snow, *From Missouri,* 78.
7. Ibid., 212.

Budge Cobb represented the temporary, but significant, victory of a simple, tough, and guileless man against forces that manipulated the land and exploited the people of the Bootheel.

In Snow's somewhat oversimplified view, Whitfield and the farm laborers who camped along the highways in 1939 shared a kinship with Budge Cobb. They were people who had a close connection to the land. Most survived, in part, by hunting and gathering. They brought their guns to the encampments and shot rabbits for food. In the eyes of passing motorists, they seemed uncouth and dangerous, but their intentions were peaceful. They wanted only the chance to make a living, to keep a roof over their heads, and to support their families. But they were tough and determined, and their quiet presence startled the comfortable and powerful—the landowners, ginners, and government officials—who had ignored their plight and relied upon their docility.

As Snow saw it, the power of the roadside demonstrators arose from several sources. On the deepest level, they derived their strength from a connection with nature. Like Snow, they were farmers. Like Budge Cobb, they were plain, uneducated people. Like all farmers and backwoodsmen, they were on familiar terms with animals, domestic and wild. Whitfield's sermons appealed to them because he spoke in understandable terms, comparing human behavior to that of mules, cattle, birds, and beasts. He assaulted the dignity of landowners and government officials by casting them in humorous barnyard roles. In this ingenious and unpretentious way, he stripped away the emotions of awe and fear that prevented farmworkers from challenging an unfair system.

Their protest had validity because they demanded only simple justice. Newspapers around the country carried photographs of the evicted families setting up housekeeping along the roadsides. Because their predicament was easily understandable and clearly wrong, readers could relate their actions to the great nonviolent movement led by Gandhi in India and to the image of Steinbeck's Okies carting their belongings to California in a fruitless quest for work.

Their protest had resonance because the world was changing. Gandhi had tested the power of passive resistance to embarrass an empire and affirm the dignity of an oppressed people. Roosevelt had promised a "new deal" not just for a few lucky Americans but indeed for every American. With all its flaws, false starts, and inconsistencies, the New Deal implanted the idea of economic entitlement. In a country with plentiful resources and uncountable wealth, no one should be homeless. No one should go hungry. The roadside demonstrators quietly and powerfully asserted their right to earn a living.

Did their protest succeed? Snow spent the last fifteen years of his life considering that question. In the aftermath of the strike, he lobbied for the reform of agricultural policy. He had some influence on the FSA's efforts to provide housing and a stable economic base for sharecroppers, but ultimately he failed, overwhelmed by changing events. Mechanization of agriculture threw more and more farm laborers off the land. The unemployment problem continued and seemed to have no remedy—until America mobilized for World War II. In that worldwide conflict, the idea of passive resistance against injustice was virtually obliterated.

Snow died at the beginning of 1955, the year after the U.S. Supreme Court banned segregation in public schools and several months before Rosa Parks challenged racial discrimination on the buses in Montgomery, Alabama. Snow would have admired Parks, a forty-two-year-old seamstress who refused to give up her seat to a white person.[8] Like the roadside demonstrators, she stood up—or rather, sat down—for simple justice. Her brave act inspired a successful bus boycott and helped to transform a Protestant minister, Dr. Martin Luther King Jr., into the leader of a national crusade for civil rights, based on the principle of nonviolent resistance to injustice.

Snow died a decade before U.S. military action in Vietnam gave rise to a nationwide antiwar movement that involved a wide array of Americans, including college students, the clergy, draft resisters, blue-collar workers, old liberals, and the New Left. While a previous generation had rationalized World War II as a "good war," in which U.S. military might triumphed over evil expansionist powers, the pacifists of the 1960s denounced the Vietnam War as the product of America's own imperialism and racism. Snow would surely have endorsed this notion. But by the end of

---

8. Maurice Isserman and Michael Kazin, *America Divided: The Civil War of the 1960s*, 29–30.

the 1960s, nonviolent demonstrations against the war frequently escalated into bloody confrontations between protesters, police, and the National Guard.[9] And by the middle 1960s, despite gains in civil rights, racial tensions erupted into violence in Los Angeles, Detroit, and elsewhere; in 1968, Martin Luther King was assassinated in Memphis, Tennessee.

At the end of his life, Snow had not given up on the possibility of peace. Despite the insane murder-suicide that robbed him of his daughters, son-in-law, and granddaughter, he believed that human beings could learn to live without violence. The roadside demonstration was a revelation for him. The peaceful protest of the sharecroppers proved to him that human beings could stand up in a dignified way, insisting on justice, without attacking their enemies. In the Cold War din of nationalistic bombast and exaggerated fear, he listened for voices of reason. By writing his essays and his book, he tried to show the world that one of those voices belonged to a lowland farmer; one of those voices came from Missouri.

Thad Snow looked out at the world from a peculiar place on the landscape. In half a century, while he watched, the countryside around him changed from wooded swampland to flat, tilled fields planted mainly in cotton. Newcomers arrived in the Missouri Bootheel, and many of them found their way to Snow's Corner, where they could stop and chat with a suntanned and bespectacled farmer out in the fields, or, in later years, with an aging man propped up in his bed, writing a book.

Thad Snow was not proud of everything he had done. He seriously questioned the wisdom of clearing the forests and draining the swamps, although he had found the process exhilarating. In the 1910s and 1920s, he pushed for concrete roads in the southeastern lowlands, but gradually he questioned the headlong rush toward highway-building and drew back from some of his booster colleagues. He enjoyed the challenge of growing cotton, but the social arrangements for cotton production troubled him. In the early 1930s, he embraced the New Deal, but he became disillusioned when he realized that the system was fraught with injustice and mismanage-

9. Ibid., 182–84.

ment. He lobbied to change the government's farm policy and sought to win a better deal for farm laborers, but world events made his efforts irrelevant. One event that gave him hope, however, was the sharecroppers' dramatic demonstration on the highways in 1939.

Snow was an unusual man in many ways. At an early age he chose to be a farmer during a time when many young, ambitious, educated people were heading for the cities. For the rest of his life, he clung to the ideal of a rural frontier, even as he helped open one of the last vestiges of the frontier to economic development. When he pushed to bring highways into the Bootheel, he behaved as a modernizer, but he distrusted modernization and made his feelings plain when he wrote nostalgically about mules and hunting and life in the backwoods.

He was ambitious and in some ways greedy. In 1910, he sought a big piece of undeveloped land on which to make a profit. In 1923, he joined the cotton boom and brought sharecroppers to his farm, and he continued to employ sharecroppers through the 1940s. He profited from their labor, but he also let them stay on as sharecroppers when most of his neighbors found it more profitable to employ seasonal day labor. Although he admitted to his selfish impulses, he also behaved in ways that undermined his own self-interest.

As a well-to-do planter, he should have been a natural opponent of unionization among farmworkers, yet he invited the STFU to organize sharecroppers on his land. A traitor to his own class, he incurred the wrath of his white landowner neighbors by standing up for the laborers, who were landless, poor, and mostly black. He endured vilification for many years and in some ways brought heartache and tragedy down on his family, but he never regretted his association with the sharecroppers' cause.

He was far from perfect. He could be a good friend, but he could also be sardonic and overbearing. In part at least, the tragedy in his family resulted from his constant needling of his son-in-law, who could not measure up to Snow's expectations. Perhaps he learned from that tragedy. The tone of his memoir was not sarcastic, but gentle, rueful, and kind.

He was a utopian, a person who dreamed of a better world. In the twentieth century, *utopian* became a pejorative term, hinting at lunacy, suggesting at the very least a lack of proper contact with reality. Many of his neighbors viewed Snow as eccentric. Others said he was just ahead of his time.[10]

The radical European philosopher Ernst Bloch described many rational forms of utopian thinking. Frontiers, by his definition, were geographic utopias, promising the potential for an earthly paradise somewhere, usually far to the west. Social utopias encompassed ideas for improving human life through wisdom and knowledge gained from the perception of present and past injustice. To envision a social utopia required an understanding of wrongs that existed and a desire to rectify them.[11]

In his adult life, Snow evolved from a geographic frontiersman to a social utopian, observing and condemning real problems but continuing to hope for solutions, not in some new and fanciful place, but right where he stood, on his home ground, and throughout the world. His utopian vision arose less from reading and meditation, although he read widely and thought deeply, than from breaking mules and hauling logs, contending with floods, pleading with the government for help during the Depression, and watching the share-croppers engage in a mighty protest. Put plainly, his vision emerged from the soil of the Bootheel.

10. For example, Millie Wallhausen said, "He was eccentric. This is the first word that comes to mind" (interview, October 28, 2002). Other interviewees, such as Hunter Rafferty and Julia Warren, said he was just thinking ahead of everyone else.
11. Ernst Bloch, *A Philosophy of the Future*, 87, 92.

Epilogue

He had a genuine love of the Ozarks.

—Fannie Snow Delaney, August 16, 1999

In the 1960s, the National Park Service (NPS) acquired more than a hundred thousand acres along the Current and Jack's Fork Rivers in the Missouri Ozarks in order to preserve the two free-flowing rivers and prevent the construction of dams. Included in this federal preserve was Big Spring, formerly a state park. The NPS set up its headquarters for the Ozark National Scenic Riverways in Van Buren, across the Current River from the site of the old Rose Cliff Hotel.

I spent some time working on a research project for the NPS in the fall of 2002. Snow's friend George Burrows had died just a few months earlier, on the fourth of July. A front-page obituary in the *Current Local* praised him as a teacher, a naturalist, and a musician. He was eighty-two. Mike Gossett, a water quality specialist for the Riverways, said George had a hard time during his last years, suffering from physical illness and depression. But he always came out to help with the recycling program in Van Buren.[1]

During my semester of research, I had some time to watch the Current River flow under the new bridge past the new hotel, called The Landing, which had replaced the Rose Cliff. It too had

1. *Van Buren Current Local,* July 11, 2002; Mike Gossett, informal conversation, September 24, 2002.

basement windows that featured views of the river. I talked to a lot of people, including NPS employees, naturalists, and long-time residents of Van Buren.

One day I was having lunch with Dr. Jim Price, the archaeologist, and Renata Culpepper, an archaeological technician. Both of them had grown up in the Ozarks, and they were talking about the way things were. It was an early fall day. The dogwood and the sumac had turned bright red, and a slight wind rustled in the partly green, partly yellow forest. We must have been thinking about the coming winter, because Renata talked about going out after the first snowstorm and cutting a jag of wood. "Why we only cut a jag—why we didn't go out and cut enough for the whole winter—I don't know."

"Because," said Doc Price, "It's not the Ozarks way. The Ozarks way is 'just enough for now.'"[2]

We laughed, because it seemed funny. It didn't seem to show a lot of sense. But later, I thought, maybe it did. Maybe it made a whole lot of sense. I remembered Snow's "big-eye," his desire to acquire the largest possible piece of land and make the largest possible profit from it, and how that led to the loss of the wilderness, which he valued, and the exploitation of workers, with whom he had sympathy. I wondered if this was why he came to the Ozarks.

It's not that simple, of course. Preserving the Current River and the woodlands surrounding it has required a massive federal effort. But there is a certain truth in this. The only way we can save any part of wilderness is to limit our desire for profit and gain, and that is also the only way we can prevent the exploitation of fellow human beings.

There is a value in the "Ozarks way" that should survive and be remembered.

2. Informal conversation, October 3, 2002.

Selected Bibliography

Manuscript Collections

Caverno, Xenophon. Papers. Western Historical Manuscript Collection, University of Missouri–Columbia.

Communist Party of the United States of America. Records of the CPUSA. Manuscript Division, Library of Congress.

Cook, Fannie. Papers. Missouri Historical Society, St. Louis.

Friant, Julien N. Papers. Regional History Collection, Southeast Missouri State University, Cape Girardeau.

Greene, Lorenzo. Papers. Western Historical Manuscript Collection, University of Missouri–Columbia.

Harper, Judge Roy. Papers. Missouri Historical Society, St. Louis.

Houck, Louis. Papers. Regional History Collection, Southeast Missouri State University, Cape Girardeau.

Moore, Joe. Scrapbook on the 1927 flood, on file at the offices of the *Charleston Enterprise-Courier,* Charleston, Mo.

Nunnelee Funeral Home. Death Records. Mississippi County, Missouri, 1910–1930. Compiled by the Mississippi County Genealogical Society, Charleston, Mo. Bound copy in Clara Drinkwater Newnam Public Library, Charleston, Mo.

O'Hare, Frank. Papers. Missouri Historical Society, St. Louis.

Snow, Thad. Papers. Western Historical Manuscript Collection, University of Missouri–St. Louis. (Microfilm on file at University Archives and Regional History Collection, Kent Library, Southeast Missouri State University, Cape Girardeau.)

Southern Tenant Farmers' Union. Papers, 1934–1970. Glen Rock, N.J.: Microfilming Corp. of America, 1971.

Stark, Governor Lloyd Crow. Papers, 1931–1941. Western Historical Manuscript Collection, University of Missouri–Columbia.

University of Missouri President's Office. Papers. Western Historical Manuscript Collection, University of Missouri–Columbia.

Watson, Goah. "Judge Goah Watson's Account of the Settlement of New Madrid." Typescript, collection 995, vol. 3, no. 1001, pp. 11–13. Western Historical Manuscript Collection, University of Missouri–Columbia.

Federal, State, and Local Records

Butler County, Missouri, Recorder. Deed books 216, 405, 546, 882, 923.

Mississippi County, Missouri, Circuit Court Record. Books 22, 26.

Mississippi County, Missouri, Recorder. Deed books 59, 63, 64, 70, 77, 79, 92, 94, 98, 100, 103, 105, 117, 125, 164.

Mississippi County, Missouri, Recorder. Marriage license book 13.

Missouri State Highway Commission. *Biennial Reports*, 1917–1932.

National Archives and Records Administration, Kansas City. RG 77. Records of the Chief of Engineers, St. Louis, Missouri, District. General Correspondence, 1926–1940.

U.S. Bureau of the Census. *Fifteenth Census of the United States: 1930.* Vol. 3, pt. 1: *Population*. Washington, D.C.: Government Printing Office, 1930.

U.S. Bureau of the Census. *Fifteenth Census of the United States: 1930. Population Schedule: Mississippi County, Missouri.*

U.S. Bureau of the Census. *Thirteenth Census of the United States: 1910.* Vol. 3, pt. 1: *Population*. Washington, D.C.: Government Printing Office, 1913.

U.S. Census, Mississippi County, 1920.

U.S. Census, Mississippi County, 1930.

U.S. Farm Security Administration. "Southeast Missouri: A Laboratory for the Cotton South." December 30, 1940.

Maps

General Map of Mississippi County, Including Parts of New Madrid and Scott County, Mo., 1920. Mississippi County Historical Society item #702.

Levee District No. 3, Mississippi County, Mo., 1925. Mississippi County Historical Society, item #701.

Map of Mississippi County, 1883. Kent Library.

Map of Mississippi County, 1918. Mississippi County Historical Society item #2478.

War Department, Corps of Engineers. Missouri–Illinois Charleston Quadrangle, 1939.

War Department, Corps of Engineers. Missouri–Illinois Charleston Quadrangle, 1954.

Newspapers

Cape Girardeau Southeast Missourian, January–March 1930; March–August 1936; January–May 1937; December 1941; August 1948; May 15, 1969.

Charleston (Mo.) Democrat, January–February 1940; December 11, 1941.

Charleston (Mo.) Enterprise-Courier, 1910–1955.

Memphis Commercial Appeal, May 1, 1927.

New York Times, November 14, 1954; January 17, 1955.

Poplar Bluff (Mo.) Daily American Republic, January–July 1939; January 17, 1955.

Scott County (Mo.) Democrat, January–March 1937.

Sikeston (Mo.) Standard, 1922–1955.

St. Louis American, February 16, 1939.

St. Louis Argus, September 19, 1952.

St. Louis Globe-Democrat, June 9, 1929; January 1, 1953.

St. Louis Post-Dispatch, May–August 1937; October 27 and 31, 1938; March 7, 1941; July 12, 1941; August 15, 1948; January 16, 1955; February 11, 2001.

Van Buren Current Local, February–August, 1924; January 20, 1955; July 11, 2002.

Oral Sources

Burrows, George. Interviews with the author, Van Buren, Mo., July 10, 1998, and December 15, 1998.

Cooper, Alex. Interviews with David Whitman, March 11, 1994, July 27, 1994, and August 13, 1994. Bootheel Project, collection 3928, audiocassettes 15 and 27. Western Historical Manuscript Collection, University of Missouri–Columbia.

———. Telephone interview with the author, June 1, 2001.

Cooper, Jennie. Telephone interview with the author, September 23, 2002.

———. Interview with the author, Cape Girardeau, Mo., October 24, 2002.

Corse, Debbie Delaney (Mrs. Wayne). Telephone interview with the author, May 28, 2002.

Cotner, Dan. Interview with the author, Cape Girardeau, Mo., April 4, 2002.

Delaney, Fannie Snow. Interviews with the author, Snow's Corner, Mississippi County, Mo., February 9, 1996, August 16, 1999, and March 19, 2001.

Farmer, Shirley Whitfield. Interview with the author, Jackson, Mo., April 7, 2002.

Fleming, Barbara Whitfield. Telephone interview with the author, September 10, 2002.

Oral history of cotton industry in Scott County, Missouri, 1880–1988. Archives and Regional History Collection, Kent Library, Southeast Missouri State University, Cape Girardeau.

Price, Acel. Interview with the author, Van Buren, Mo., September 12, 2002.

Rafferty, Hunter. Telephone interview with the author, May 28, 2002.

Robinson, James. Telephone interview with the author, September 10, 2002.

Stallings, Nellie Feezor. Interview with the author, Charleston, Mo., April 1, 2001.

Wallhausen, Millie. Interview with Ray Brassieur, June 14, 1996, Charleston, Mo. Bootheel Project, collection 3965, audiocassette 12. Western Historical Manuscript Collection, University of Missouri–Columbia.

———. Interview with the author, Charleston, Mo., October 28, 2002.

Warren, Julia Cooper. Interview with the author, Charleston, Mo., November 7, 2002.

References

"Battle of Belmont." Special publication of the *Charleston (Mo.) Enterprise-Courier,* October 1991.

"Beckwith Collection: An Illustrated Vignette of the Mound Builders, with Photographs." Cape Girardeau: Southeast Missouri State University Museum, n.d., n.p.

Belfrage, Cedric. "Cotton-Patch Moses." *Harper's Magazine* 36 (November 1948): 94–103.

Benton, Thomas Hart. *An Artist in America.* 4th ed. Columbia: University of Missouri Press, 1983.

Berry, Wendell. *Life Is a Miracle: An Essay against Modern Superstition.* Washington, D.C.: Counterpoint, 2000.

———. *The Unsettling of America: Culture and Agriculture.* San Francisco: Sierra Club Books, 1977.

Berthe, L. T. *Old Man River Speaks.* Charleston, Mo.: Berthe, 1937.

Bloch, Ernst. *A Philosophy of the Future.* New York: Herder and Herder, 1970.

Bowen, B. F. *Biographical Memoirs of Hancock County, Indiana.* Logansport, Ind.: B. F. Bowen, 1902.

Bratton, Sam T. "Land Utilization in the St. Francis Basin." *Economic Geography* 6 (1930): 374–88.

Brock, Peter. *Pacifism in the United States, from the Colonial Era to the First World War.* Princeton, N.J.: Princeton University Press, 1968.

———. *Twentieth-Century Pacifism.* New York: Van Nostrand Reinhold, 1970.

Buckingham, Peter H. *Rebel against Injustice: The Life of Frank P. O'Hare.* Columbia: University of Missouri Press, 1996.

Cadle, Jean Douglas. "Cropperville, from Refuge to Community: A Study of Missouri Sharecroppers Who Found an Alternative to the Sharecropper System." Master's thesis, University of Missouri–St. Louis, 1993.

Campbell, Rex R., John C. Spencer, and Ravindra G. Amonker. "The Reported and Unreported Missouri Ozarks: Adaptive Strategies of the People Left Behind." In *Forgotten Places: Uneven Development in Rural America,* ed. Thomas A. Lyson and William W. Falk, 30–52. Lawrence: University Press of Kansas, 1993.

Cantor, Louis. *A Prologue to the Protest Movement: The Missouri Sharecropper Roadside Demonstration of 1939.* Durham, N.C.: Duke University Press, 1969.

Capeci, Dominic J., Jr. *The Lynching of Cleo Wright.* Lexington: University of Kentucky Press, 1998.

Chambers, Henry E. *Mississippi Valley Beginnings: An Outline of the Early History of the Earlier West.* New York: G. P. Putnam's Sons, 1922.

Chapman, Carl H., and Eleanor F. Chapman. *Indians and Archaeology of Missouri.* Columbia: University of Missouri Press, 1983.

Chase, Stuart. "From the Lower Depths." *Reader's Digest* 38 (May 1941): 108–12.

Christensen, Lawrence O., William E. Foley, Gary R. Kremer, and Kenneth H. Winn, eds. *Dictionary of Missouri Biography.* Columbia: University of Missouri Press, 1999.

Conrad, David Eugene. *The Forgotten Farmers: The Story of Sharecroppers in the New Deal.* Urbana: University of Illinois Press, 1965.

Conzen, Michael P., ed. *The Making of the American Landscape.* Winchester, Mass.: Unwin Hyman, 1990.

Cook, Fannie. *Boot-heel Doctor.* New York: Dodd, Mead, 1941.

Daniel, Pete. *The 1927 Mississipi River Flood.* New York: Oxford University Press, 1977.

Dennis, Lawrence. *The Dynamics of War and Revolution.* New York: Weekly Foreign Letter, 1940.

Doherty, William T., Jr. *Louis Houck: Missouri Historian and Entrepreneur.* Columbia: University of Missouri Press, 1960.

Donovan, Robert J. *Tumultuous Years: The Presidency of Harry S. Truman, 1949–1953.* New York: W. W. Norton, 1982.

Feezor, Joan Tinsley. "Marcus Alphonse Murphy and Communism on Trial: The Smith Act in Missouri." Master's thesis, Southeast Missouri State University, 1993.

Finiels, Nicolas de. *An Account of Upper Louisiana.* Ed. Carl J. Ekberg and William E. Foley. Columbia: University of Missouri Press, 1989.

Forister, Robert H. *Complete History of Butler County, Missouri*. Marble Hill: Stewart Publishing, 1999.

Goldman, Helen Frances Levin. "Parallel Portraits: An Exploration of Racial Issues in the Art and Activism of Fannie Frank Cook." Ph.D. diss., St. Louis University, 1992.

Goodspeed's History of Southeast Missouri. 1888. Reprint, Cape Girardeau: Ramfre Press, 1964.

Grantham, Dewey W. *The South in Modern America*. New York: HarperCollins, 1994.

Greene, Lorenzo J. "Lincoln University's Involvement with the Sharecropper Demonstration in Southeast Missouri, 1939–1940." *Missouri Historical Review* 82 (October 1987): 24–50.

Greene, Lorenzo J., Gary R. Kremer, and Antonio F. Holland. *Missouri's Black Heritage*. Columbia: University of Missouri Press, 1993.

Grubbs, Donald H. *Cry from the Cotton: The Southern Tenant Farmers' Union and the New Deal*. Chapel Hill: University of North Carolina Press, 1971.

Hall, Leonard. *Stars Upstream: Life along an Ozark River*. Columbia: University of Missouri Press, 1969.

History and Families, Mississippi County, Missouri, 1845–1995. Paducah, Ky.: Turner Publishing, 1995.

Hobson, Fred. *But Now I See: The White Southern Racial Conversion Narrative*. Baton Rouge: Louisiana State University Press, 1999.

Holley, Donald. *Uncle Sam's Farmers: The New Deal Communities in the Lower Mississippi Valley*. Urbana: University of Illinois Press, 1975.

Houck, Louis. *A History of Missouri from Earliest Explorations and Settlements until the Admission of the State into the Union*. 3 vols. Chicago: R. R. Donnelley and Sons, 1908.

Howe, Irving. *Leon Trotsky*. New York: Viking, 1978.

Isserman, Maurice, and Michael Kazin. *America Divided: The Civil War of the 1960s*. New York: Oxford University Press, 2000.

Kester, Howard. *Revolt among the Sharecroppers*. Knoxville: University of Tennessee Press, 1997.

Limerick, Patricia Nelson. *The Legacy of Conquest: The Unbroken Past of the American West*. New York: Norton, 1987.

Little River Drainage District of Southeast Missouri, 1907–Present. Cape Girardeau, Mo.: Little River, 1989.

Lowe, Jimmy. *Striving Upward: An Autobiography.* Mountain Home, Ark.: James L. Lowe, 1996.

McWilliams, Carey. *Ill Fares the Land: Migrants and Migratory Labor in the United States.* Boston: Little, Brown, 1942.

Missouri: The WPA Guide to the "Show Me" State. Compiled by the Workers of the Writers' Program of the Works Projects Administration in the State of Missouri. 1941. Reprint, St. Louis: Missouri Historical Society Press, 1998.

Mitchell, H. L. *Mean Things Happening in This Land.* Montclair, N.J.: Allenheld, Osmun, 1979.

Mitchell, Steve. "Homeless, Homeless Are We . . ." *Preservation Issues* (Missouri Department of Natural Resources, Historic Preservation Program) 3 (January/February 1993): 1, 9–10.

Morrow, Lynn. "Rose Cliff Hotel: A Missouri Forum for Environmental Policy." *Gateway Heritage* 3, no. 2 (1982): 38–48.

Ogilvie, Leon Parker. "Governmental Efforts at Reclamation in the Southeast Missouri Lowlands." *Missouri Historical Review* 64 (January 1970): 151–76.

———. "Populism and Socialism in the Southeast Missouri Lowlands." *Missouri Historical Review* 65 (January 1971): 159–83.

Penick, James Lal. *The New Madrid Earthquakes.* Rev. ed. Columbia: University of Missouri Press, 1981.

Percy, William Alexander. *Lanterns on the Levee.* 1941. Reprint, Baton Rouge: Louisiana State University Press, 1973.

Pisani, Donald J. "Beyond the Hundredth Meridian: Nationalizing the History of Water in the United States." *Environmental History* 5, no. 4 (2000): 467–82.

Powell, Betty F. *History of Mississippi County, Missouri, Beginning through 1972.* Independence, Mo.: BNL Library Service, 1975.

Pressly, Thomas J., and William Scofield, eds. *Farm Real Estate Values in the United States by Counties, 1850–1959.* Seattle: University of Washington Press, 1965.

Revell, Peter. *James Whitcomb Riley.* New York: Twayne Publishers, 1970.

Richman, George J. *History of Hancock County, Indiana.* Indianapolis: Federal Publishing, 1916.

Ridpath, Ben. "The Case of the Missouri Sharecroppers." *Christian Century* 56 (February 1, 1939): 146–48.

Riley, James Whitcomb. *The Complete Poetical Works of James Whitcomb Riley.* New York: Garden City Publishing, 1941.

———. *Riley Farm-Rhymes.* Indianapolis: Bobbs-Merrill, 1921.

Salvatore, Nick. *Eugene V. Debs, Citizen and Socialist.* Urbana: University of Illinois Press, 1982.

Sarvis, Will. "Black Electoral Power in the Missouri Bootheel, 1920s–1960s." *Missouri Historical Review* 95 (January 2001): 182–202.

Schott, Webster. "Thad Snow, the Farmer." *New Republic* 12 (April 18, 1955): 16–17.

Schroeder, Rebecca B., and Donald M. Lance. "John L. Handcox, 'The Sharecropper Troubador.' " *Missouri Folklore Society Journal* 8–9 (1986–87): 123–42.

Shrum, Edison. "Super Floods Raging in Wide Spread Area: The *Scott County Democrat*'s Account of the 1937 Mississippi River–New Madrid County Jadwin Floodway Disaster," with additional text by Edison Shrum. Scott City, Mo., 1994. Typescript.

Snow, Thad. *From Missouri.* Boston: Houghton Mifflin, 1954.

———. "Proud Kate, the Aristocratic Mule." *Harper's Magazine* 209 (July 1954): 64–68.

———. "When Traders Rule, Ruin Follows." *Missouri Farmer,* December 15, 1931, 13.

———. "Why Share-Croppers Join the CIO." *St. Louis Post-Dispatch,* August 9, 1937.

Steinbeck, John. *The Grapes of Wrath.* 1939. Reprint, New York: Bantam, 1972.

———. "The Leader of the People." In *The Long Valley.* 1938. Reprint, New York: Penguin Books, 1995.

Stepenoff, Bonnie. "Cotton Comes to the Bootheel: Thad Snow and Social Change in Southeast Missouri, 1923–1939." *Red River Valley Historical Journal* 2 (fall 2002): 37–57.

———. "The Last Tree Cut Down: The End of the Bootheel Frontier." *Missouri Historical Review* 90 (October 1995): 61–78.

———. "Mother and Teacher as Missouri State Penitentiary Inmates: Goldman and O'Hare, 1917–1920." *Missouri Historical Review* 85 (July 1991): 402–21.

Streeter, Jean Douglas. "The Fannie Cook Papers." *Gateway Heritage* 9 (winter 1988–1989): 43.

Strickland, Arvarh E. "The Plight of the People in the Sharecroppers' Demonstration in Southeast Missouri." *Missouri Historical Review* 81 (July 1987): 403–16.

Thelen, David. *Paths of Resistance: Tradition and Democracy in Industrializing Missouri.* Columbia: University of Missouri Press, 1986.

Theoharis, Athan. *Seeds of Repression: Harry S. Truman and the Origins of McCarthyism.* Chicago: Quadrangle Books, 1971.

Turner, Frederick Jackson. *The Frontier in American History.* New York: Henry Holt, 1920.

Vance, Rupert. *Human Factors in Cotton Culture: A Study in the Social Geography of the American South.* Chapel Hill: University of North Carolina Press, 1929.

Veblen, Thorstein. *Theory of the Leisure Class.* New York: Penguin, 1994.

Walsh, J. Omer. *Boots and Walsh's Directory of the City of Greenfield, 1893–1894.* Greenfield, Ind.: Walsh, 1894.

White, Max R., and Douglas Ensminger. *Rich Land—Poor People.* United States Department of Agriculture, Farm Security Administration. Washington, D.C.: Government Printing Office, 1938.

Wittner, Lawrence S. *Rebels against War: The American Peace Movement, 1941–1960.* New York: Columbia University Press, 1969.

Wyllie, Irvin J. "Race and Class Conflict on Missouri's Cotton Frontier." *Journal of Southern History* 20 (May 1954): 183–95.

Index

Acheson, Dean, 148–51
African Americans: migration to the Bootheel, 14, 20–21, 54, 57–58; on Snow's farm, 20, 38, 60, 66–70, 77–78, 96–97, 143; as sharecroppers, 57–59, 70, 90; and Southern Tenant Farmers' Union (STFU), 72, 75–77; participation in Sharecroppers' Roadside Demonstration, 92, 96; and Fellowship of Reconciliation, 136. *See also* Civil rights; Cropperville; La Forge, Mo.; Lynching; Racial conflict; Racial segregation
Agricultural Adjustment Act (AAA), 61–62, 66–67, 72, 74, 81, 111
Alabama, 71, 152, 159
Alfalfa, 15, 28, 54
American Civil Liberties Union (ACLU), 113
American Federation of Labor (AFL), 76
American Friends Service Committee, 110
American Red Cross, 43–44, 50, 95, 116–17
American Revolution, 10
Americans for Democratic Action, 129
Anderson, Victor E., 66–67
Anderson, William R., 79

Anti-Semitism, 7, 121, 133
Appeal to Reason, 74
Arkansas, x, 5, 26, 44, 54–55, 64, 72–77, 82, 144
Armstrong, Sam, 90–91, 102
Associated Farm Laborers, Sharecroppers, and Tenant Farmers of Southeast Missouri (AFLST), 119

Baasch, Hans, 64, 82, 92, 100, 102–3
Baker, Charles B., 21
Bank of Sikeston, 16, 103
Bayouville, Mo., 49
Beck, Walter, 68
Beckwith, Thomas, 10
Belfrage, Cedric, 81, 90–91
Belmont, Mo., 12, 49
Benton, Mo., 50
Benton, Thomas Hart, 39
Berry, Wendell, 22, 25
Berthe, Lucius T., 45, 51
Big Lake, 42–43, 51
Big Spring State Park, x, 145, 163
Bird, Abraham, 39
Bird's Point, Mo., 9, 32, 40, 46
Blanton, Charles L., 94, 100–104, 123
Blanton, David L., 123
Bloch, Ernst, 162
Bollinger County, Mo., 15
Boll weevil, 55–56, 69
Boone, Daniel, 23–24

Bootheel: Great Depression in, 8,
    68–70; early history of, 10–13;
    population boom in, 20–21,
    53–57, 60–64; mentioned, 138,
    143, 147–48, 151, 153, 158, 160,
    162. *See also* Floods; Flood con-
    trol; Highways; Sharecroppers'
    Roadside Demonstration;
    Swamp drainage; Swampeast
    Missouri
Boot-heel Doctor (Fannie Cook), 106,
    116–18, 120, 122, 124
Bryan, William Jennings, 128
Burgess, David, 113
Burrows, Emmet Russell "Rip," ix, xi,
    138, 163
Burrows, George, ix–xiv, 138, 144
Butler, J. R., 91, 95
Butler, Pauline Hawkins. *See* Murphy,
    Pauline Hawkins Butler
Butler County, Mo., 6, 108–10

Cairo, Ill., 9, 29, 32, 36, 44–47, 50, 92,
    140, 152, 156
Canalou, Mo., 15
Cape Girardeau, Mo., 10–11, 13, 15,
    33, 36, 44, 50, 105, 114, 130
Cardenas, Lazaro, 84–85
Carter County, Mo., 144, 146
Caruthersville, Mo., 21, 53
Casteel, B. M. (Colonel), 91–92, 94–95
Catt, Alma, 17, 142
Catt, Lawrence, 17
Caverno, Xenophon, 15–17, 23, 31, 89,
    104, 123
Centennial Road Law, 32
Chambers, Whittaker, 148–49
Charleston, Mo., ix, 2, 4, 6, 9–10, 18,
    21, 29, 32–33, 42, 48, 50, 53,
    67–68, 86, 90, 92, 95–96, 98, 104,
    107, 111, 141, 143, 151–52, 155–56
Charleston Democrat, 130
Charleston Enterprise-Courier, 21, 104,
    156
Charleston High School, 20, 98
Circle City, Mo., 112
Civilian Conservation Corps (CCC),
    50, 144
Civil rights, 125, 159
Civil War, 12, 26, 54, 124

Clark, Bennett Champ, 100
Clark National Forest, 145
Class divisions, 16, 19–22, 35–37, 38,
    84–85
Coast Guard, 44
Cobb, Budge, 157–58
Coghlan, Ralph, 122
Cold War, 148–50, 160
Communism, 72, 121, 125, 148–51
Communist Party of the U.S.A.
    (CPUSA), 71–72, 79, 148, 151–52
Concord, Mo., 22, 142
Congress of Industrial Organizations
    (CIO), 76, 79, 83, 94
Cook, Fannie, 94, 99, 106–7, 110,
    115–23
Cooper, Alex, 59–60, 69, 143–44
Cooper, Anna D., 16
Cooper, Jennie, 2
Cooter, Mo., 58
Corn, 28, 35, 53–54
Corps of Engineers, 44–48, 51–52,
    157. *See also* Cobb, Budge
Corse, Debbie, 144
Corse, Wayne, 144
Cotton: in the Bootheel, 4–5, 16,
    53–59, 69–70, 160; plantation sys-
    tem, 54, 56, 59–60; and Great
    Depression, 61, 65; and farm
    mechanization, 82–83, 119, 144;
    mentioned, 35, 89. *See also*
    Agricultural Adjustment Act
    (AAA); Boll weevil; Planters;
    Sharecroppers; Tenant farmers
Courtois Hills, 144
Cropperville, 6–7, 36, 108–10, 114, 144
Crosno, Mo., 49
Culpepper, Renata, 164
Current Local, ix, xiii, 163
Current River, ix–x, 145–46, 163–64
Currin, Marshal. *See* Kern, Marshal
Cutler, T. H., 33
Cypress trees, 9, 34, 37

Davis, Ben, 138, 145–46
Davis, Robert, 145
Debs, Eugene V., 74, 129
Deering, Mo., 14, 21
Deforestation, 14–15, 21, 160. *See also*
    Lumber industry

Delaney, Bob, xi, xiii, 8, 22, 143
Delaney, Fannie Snow, xi, xiii, 3, 7–8,
    18, 20, 22, 31, 34–35, 104, 122,
    125, 136, 139, 143–44, 153, 163
Delmo Housing Corporation, 113
Delmo Project, 112–13
Delta of the Mississippi River, 9, 39,
    72. *See also* Mississippi River
Democratic Party, 17, 29, 132, 151
Denton, Mo., 58
Depression. *See* Great Depression
Dexter, Mo., 13, 141
Dorena, Mo., 44, 49, 92, 98
Drainage. *See* Swamp drainage
Dudley, Harry E. (Colonel), 89, 96–97
Dunklin County, Mo., 12–13, 15,
    55–57, 69, 100
Du Quoin, Ill., 36, 114

East, Clay, 73
East Prairie, Mo., 50, 106
Emerson, Jim Mac, 58
England, 128, 132–33
Environmentalism, 25, 37–38

Farmer, Shirley Whitfield, 6–7, 81,
    102, 109–10, 114
Farmers Home Administration
    (FHA), 113
Farmers' Institute, 9, 28–29
Farmington, Mo., 141
Farm Security Administration (FSA),
    62–64, 106, 109, 111–13, 119, 159
Farmworkers. *See* Sharecroppers;
    Tenant farmers
Fascism, 120–21, 125
Federal Bureau of Investigation (FBI),
    xiv, 100, 151–52
Federal Emergency Relief
    Administration (FERA), xiv, 100
Federal housing programs, 111–12.
    See also Delmo Project; La Forge,
    Mo.; Southeastern Missouri
    Scattered Labor Homes Program
Feezor, Mose, 19–20, 22, 37–38
Feezor, Nellie. *See* Stallings, Nellie
    Feezor
Fellowship of Reconciliation (FOR),
    99, 136
Finiels, Nicolas de, 11

Fischer, William R., 98, 108
Fitzpatrick, Dan, xi, 154
Fleming, Barbara Whitfield, 102, 155
Flood control, 29, 40–41, 44–47
Floods: of 1912, 40–43; of 1913, 43; of
    1927, 43–44; of 1937, 47–51, 157;
    mentioned, 4, 29, 39, 69, 117. *See
    also* Mississippi River
France, 132
Friant, Julien, 61, 67
From Missouri (Thad Snow), 8, 39, 52,
    93, 103, 126, 155–58
Frontier, 4, 8–9, 24–25, 83, 153–54,
    160. *See also* Pioneers

Gandhi, Mahatma, xiii, 125–27,
    135–36, 154, 158
Gellhorn, Edna, 94, 106
Gideon, Mo., 14
Gilmore, Ernest G., 106
Gilmore, Maude, 106–7
Gossett, Mike, 163
Gravel Ridge, Mo., 22
Great Depression, xii, 4–5, 34–35, 37,
    47, 60–66, 69, 76, 83, 125, 134,
    138, 144–45, 152, 162. *See also*
    New Deal
Greene, Lorenzo J., 98–99
Greenfield, Ind., 2, 26–28
Grigsby School, 19

Hall, Leonard, xi, 138, 145–46, 156–57
Hancock County, Ind., 9, 15, 25, 28
Handcox, John L., 5, 8, 71, 73, 76–79,
    87–88
Harvard University, 149
Harviell, Mo., 6, 110
Hatcher, Robert A., 12
Hayti, Mo., 92, 98
Henderson, Donald, 79
Highways: in the Bootheel, 6, 20,
    160–61; promoted by Thad
    Snow, 32–34. *See also* Missouri
    State Highway Commission;
    Missouri State Highway
    Department; Missouri State
    Highway Patrol; Sharecroppers'
    Roadside Demonstration
Himmelberger and Harrison Lumber,
    13

Hiss, Alger, 148–49
Hitler, Adolf, 7, 121, 133
Holland, Mo., 58
Homeless Junction, 98, 105
Houck, Louis, 12–13
House Un-American Activities
    Committee (HUAC), 148–49
Hull, Cordell, 105
Hunting, xii, 3, 14, 24

Illinois, 13, 40, 130
Imperialism, 132, 159
India, 128
Indiana, 2, 9–10, 17, 26, 28–30
Indianapolis, Ind., 26
International Harvester, 14, 21
Italy, 130

Jack's Fork River, 145, 163
Jackson, Adah, 2
Jackson, Bess. See Snow, Bess Jackson
Jackson, Solomon, 2
Jadwin, Edgar (Major General),
    44–45, 48
Japan, 126–27; 129–30; 132–33, 137
Jefferson City, Mo., 90, 98
Johnson, Josephine, 99
Johnson, Walter, 98
Jonesboro, Ark., x

Kennett, Mo., 13, 90
Kentucky, 12, 19, 47, 68
Kern, Marshal, 96–97
Kester, Howard, 72, 74–75, 79, 82
King, Austin A. (Governor), 12
King, Martin Luther, Jr., 127, 159–60
Kirkwood, Mo., 102, 108
Korean War, xi, 151–52
Ku Klux Klan (KKK), 101–2

Labor unions. See American
    Federation of Labor (AFL);
    Associated Farm Laborers,
    Sharecroppers, and Tenant
    Farmers of Southeast Missouri
    (AFLST); Congress of Industrial
    Organizations (CIO); Share
    Croppers Union (SCU); Southern
    Tenant Farmers' Union (STFU);
    United Cannery, Agriculture,

Packing, and Allied Workers of
    America (UCAPAWA)
La Forge, Mo., 64–65; 81–82, 87, 92,
    94, 98–99, 101–2
Landowners, 59–60, 62, 86–87, 89,
    100, 104, 110–11, 115, 122, 158,
    161. See also Planters
Lane, Tomy, 58–59
Levees. See Flood control
Lewis, John L., 76, 79
Lincoln University, 98–99
Lister, Jake, 146–47
Little River Drainage District, 15. See
    also Swamp drainage
Louisiana, 44, 47, 55, 74
Lumber industry, 12–14, 21. See also
    Deforestation
Lynching, 21, 118, 122–23

Malden, Mo., 56, 107
Manker, Charles, 112
Manker, Florence, 112
Manker, Helen, 112
Manker, James, 112, 141
Marked Tree, Ark., 74
Martin, Albert, 106
Matthews, Charles D., 16–17, 32–33
Matthews, Joe L., 103
Matthews, Mo., 99
Mattox, Cleve, 93
McCarthy, Joseph, 151–52
McCarthyism, 151–53
Medley, Mo., 49
Memphis, Tenn., 31, 48, 74, 76, 79, 99,
    160
Mexican War, 127
Mexico, 84–85, 89, 104–5
Mississippi, 29, 44, 58, 73
Mississippians, 10
Mississippi County, Mo., xiii, 2–3, 5,
    9, 12, 15, 17, 29, 32, 39–46, 55–57,
    66–67, 69, 80, 89, 94, 96, 111, 113
Mississippi River, 4, 9–11, 29, 31, 39,
    43, 45, 47, 51, 53, 72, 92.
    See also Delta of the Mississippi
    River; Flood control; Floods
Mississippi River Commission, 43
Missouri Committee for the
    Rehabilitation of the
    Sharecroppers. See St. Louis

Committee for the Rehabilitation
of the Sharecroppers
Missouri Conservation Commission,
145
Missouri National Guard, 89, 96
Missouri State Employment Service,
111
Missouri State Highway
Commission, 32–33, 95
Missouri State Highway Department,
33, 97
Missouri State Highway Patrol, 6,
91–92, 95–96
Mitchell, H. L., 73–74, 76–79, 95, 106
Moore, Joe, 95
Morgan, George (Colonel), 10–11
Morley, Mo., 58
Mounds, Ill., 114
Mounds (Indian), 10, 36
Mules, 18–20, 28, 158, 160
Murphy, Marcus Alphonse ("Al"),
71–72, 151–54
Murphy, Pauline Hawkins Butler, 152
Muste, A. J., 136

Napoleon, 133
National Labor Relations Act, 76
National Park Service (NPS), 144,
163–64
National Planning Association,
134–35
Nazism, 121. See also Fascism; Hitler,
Adolf
New Deal: farm recovery programs
of, 5, 61–64, 67, 84–85; and World
War II, 131–32; mentioned, 17,
72, 102, 148, 160. See also
Roosevelt, Franklin D.
New Left, 159
New Madrid, Mo., 11, 44–45, 50
New Madrid County, Mo., 12, 15, 40,
55–57, 64, 98, 100, 105, 119
New Madrid earthquake, 11
New Madrid Floodway, 40–41, 44–46,
48–49
New Orleans, La., 53
Nixon, Richard M., 149
Nonviolence, 127–28, 135–36, 159–
60
North, T. J., 96

North Atlantic Treaty Organization,
149
North Carolina, 56
Nunnelee, John F., Jr., 140–41

O'Hare, Frank P., 129, 138–39, 143
O'Hare, Kate Richards, 74, 129
Ohio River, 10–11, 29, 43, 45, 47
Old Man River Speaks (L. T. Berthe), 51
Ozark National Scenic Riverways,
145, 163
Ozarks, ix–xi, 38–39, 109, 138, 144–47,
150, 153–54, 156, 163–64. See also
Van Buren, Mo.

Parker, Harry F., 95
Parks, Rosa, 159
Peach Orchard, Mo., 119
Pearl Harbor, 125–26; 129–30, 132
Pemiscot County, Mo., 12, 14, 15, 21,
55–57, 69, 100, 119
Pioneers, 3, 23–24, 30, 38. See also
Frontier
Planters, 59, 61–62, 66, 69, 75–76, 82,
87–88, 90, 94, 99–100, 105. See also
Landowners
Poinsett County, Ark., 74–75
Poplar Bluff, Mo., 32, 93, 108, 119, 144
Portageville, Mo., 33
Price, Jim, 164
Progressive Party, 17

Quail, xii, 24
Quakers, 37, 109–10

Racial conflict, 21, 58, 65, 122. See also
African Americans; Lynching,
Vigilantes
Racial segregation, 14, 20, 50, 64, 108.
See also African Americans
Rafferty, Hunter, ix, xiii–xiv, 2, 69
Railroads, 12–13
Rankin, Jeannette, 126
Rat, Mo., 146–47
Reed, Jake, 70, 96
Reeves, Odie, 94
Republican Party, 17, 26, 34–35, 123,
148, 151
Rich Land—Poor People (White and
Ensminger), 62–63

Riggly, Orange, 106
Riley, James Whitcomb, 27–28
Road building. *See* Highways
Robinson, Jim, xiii–xiv
Rodgers, Ward, 74–75
Roosevelt, Eleanor, 108
Roosevelt, Franklin D., 61–62, 76, 85,
    105, 108, 131, 133, 148, 150, 158–59
Rose Cliff Hotel, ix–xi, 8, 145–46,
    154–55, 163. *See also* Davis, Ben;
    Davis, Robert; Van Buren, Mo.
Ross, Charles G., 68, 84

Salvation Army, 50
Scott County, Mo., 12, 15, 29, 32, 45,
    55–58, 100
Scott County Democrat, 50
Scott County Kicker, 74
Sentner, William, 152
Sharecroppers: on Snow's farm, xii,
    xiii, 5, 7, 60, 66–70, 77–78, 96–97,
    106, 161; living conditions of,
    57–59, 63–64, 66–69, 98, 106–7;
    and Great Depression, 60, 62–70;
    eviction of, 72, 76, 84, 87–89,
    91–93, 97, 105–6, 111; mentioned,
    4–5, 21–22, 49. *See also*
    Cropperville; La Forge, Mo.;
    Tenant farmers
Sharecroppers Camp. *See*
    Cropperville
Sharecroppers' Roadside
    Demonstration, xii, 6, 83, 89,
    91–98, 135, 157–59. *See also*
    Homeless Junction; St. Louis
    Committee for the Rehabilitation
    of the Sharecroppers; Sweet
    Home Baptist Church; Whitfield,
    Owen
Share Croppers Union (SCU), 71–72,
    152
Sharecroppers Voice, The, 77
Shrum, Edison, 45, 47–48
Sierra Club, xii
Sikeston, Mo., 6, 9, 53, 90, 92, 102,
    122–23
Sikeston Standard, 53, 91–95, 98,
    100–106, 115, 119, 123, 130
Simpson, Fannie, 3
Simpson, H. G. "Chilli," 33, 95–97

Simpson, John L., 3
Simpson, Lawrence A., 34, 153
Simpson, Lila. *See* Snow, Lila
    Simpson
Smith, Al, 105
Smith Act, 151–52
Snakes, 147–48
Snow, Bess Jackson, 2, 17, 26, 28
Snow, Emily: death of, 7–8, 26,
    140–43; and her father, 84, 124,
    126–27, 129, 138–40; mentioned,
    3, 20, 35, 105, 136
Snow, Hal, 2, 17, 20, 124, 139, 143, 153
Snow, Henry, 18, 25–27
Snow, Lena, 26
Snow, Lena Frances. *See* Delaney,
    Fannie Snow
Snow, Lila Simpson, 3–4, 18, 26,
    34–35, 50, 124
Snow, Mary Ann Monks, 25
Snow, Priscilla. *See* Thompson,
    Priscilla Snow
Snow, Ralph, 18, 26
Snow, Sarah Frances (Fannie)
    Pierson, 18, 24–26
Snow, Thad: character and personal-
    ity of, ix, xiv, 1–2, 19–20, 27,
    35–37, 121–23, 161; political phi-
    losophy of, x–xi, 7, 17, 34, 38, 52,
    72, 83, 86, 125, 133–37, 149–50,
    161–62; in the Ozarks, x–xi, 8, 38,
    146–47; pacifism of, xi–xii, 7,
    125–27, 130–37; as a hunter, xii,
    24; concern for sharecroppers,
    xii–xiii, 66–70, 77–78, 110–11;
    death of, xiii, 8, 156; enthusiasm
    for Thorstein Veblen, xiii, 35, 81,
    85–86; as a farmer, 2, 4–5, 15–16,
    18–19, 28–29, 35, 37, 53–54,
    83–85, 160; life in Indiana, 2,
    9–10, 17, 25–28; childhood of, 2,
    26–28; move to Missouri, 2–4,
    9–10, 15, 23–25, 29–30; as a fam-
    ily man, 3, 7, 17–18, 139; and
    John L. Handcox, 5, 8, 77–79; and
    Owen Whitfield, 5, 36, 78–83,
    89–90, 94, 100–101, 103–5, 114,
    122, 124–25, 135, 144; and
    Sharecroppers' Roadside
    Demonstration, 6–7, 89–97,

100–105, 135, 157–61; landholdings of, 16, 31, 34, 61; boosterism
of, 16, 31–32, 35; and farm policy,
17, 61–62, 131, 159, 161; paternalism of, 19–20, 22, 60, 69–70, 80,
85–86, 116, 123; as "Devil of the
Bootheel," 37, 103–4, 137;
accused of cattle rustling, 43; and
Corps of Engineers, 46, 51–52,
157; health problems of, 50, 81,
84, 155–56; and New Deal, 61,
65–66, 84, 132, 160–61; and
racism, 65, 122–23, 132–33; as
traitor to his class, 65–66, 72,
84–85, 115, 124; and labor
unions, 77–83, 86–87, 161; in
Mexico, 84, 89; and daughter
Emily, 84, 124, 129, 138–40, 143.
See also From Missouri; "True
Confession"
Snow, Thomas P., 25
Snow's Corner, xi, 2, 8, 18–19, 22, 43,
60, 77, 81, 114, 138–40, 143–44,
147, 160
Socialist Party, 17, 77, 129
South Africa, 128
South Carolina, 56
Southeastern Missouri Scattered
Labor Homes Program, 112
Southeast Missouri. *See* Bootheel;
Swampeast Missouri
Southeast Missouri Agricultural
Bureau, 31
Southeast Missouri Farms. *See* La
Forge, Mo.
Southeast Missouri State University,
10
Southern Tenant Farmers' Union
(STFU), 5, 72–79; 82–83, 86, 89,
91, 94–95, 100, 106, 119, 161
Soviet Union, 148–49, 152
Spain, 10–11
Stalin, Joseph, 150
Stallings, Marshall, 22
Stallings, Nellie Feezor, ix, 19–20, 22,
86
Stark, Lloyd C. (Governor), 68, 89–91,
95–97, 100, 103
Steinbeck, John, 153–54, 158
St. Francis River, 11, 47

St. Louis, Mo., 13, 71, 94, 99, 101, 108,
113, 129, 145, 152, 155
St. Louis American, 99
St. Louis Committee for the
Rehabilitation of the
Sharecroppers, 6, 99, 108, 152. *See
also* Cropperville
St. Louis Post-Dispatch, 5, 7–8, 24, 68,
79–80, 84–85, 90, 95, 103, 121,
129, 136, 145, 154
St. Louis Rip-Saw, 74
St. Louis Urban League, 99
Stoddard County, Mo., 12–13, 15,
55–57, 112, 141
Sunset Addition (Sikeston, Mo.), 90,
122–23
Swamp drainage, 11–12, 15, 30–31,
40, 160
Swampeast Missouri, 8, 20, 30, 124
Swank, Ruel, 141
Sweet Home Baptist Church, 98

Tenant farmers, 5, 21–22, 57–59, 62,
71–72, 84. *See also* Sharecroppers
Tennessee, 12, 64
Texas, 55, 84, 126
Theory of the Leisure Class, The
(Thorstein Veblen), xiii, 35–37
Thomas, Norman, 62, 74–75, 105, 129
Thompson, Ann, 7–8, 140–42
Thompson, John Hartwell, 7–8, 137,
140–43
Thompson, Priscilla Snow, 2, 7–8, 17,
20, 26, 124, 137, 139–42
Thoreau, Henry David, 127–28
Tolstoy, Leo, 127–28, 136
Treaty of Versailles, 131
Trotsky, Leon, 7, 104–5
"True Confession" (Thad Snow), 7,
104–5
Truman, Harry S., 100, 148–51
Tucker, Wade, 118–20
Turley, Alan, ix
Turner, Frederick Jackson, 24–25
Tyronza, Ark., 72

United Cannery, Agriculture,
Packing, and Allied Workers of
America (UCAPAWA), 79, 83, 95,
99

United States Army Corps of
   Engineers. *See* Corps of
   Engineers
United States Coast Guard, 44
United States Department of
   Agriculture (USDA), 5, 62, 82,
   111, 131. *See also* Farm Security
   Administration (FSA)
United States Forest Service, 144, 146
University of Michigan, 28
University of Wisconsin, 138–39

Van Buren, Mo., ix–xiii, 138, 144–45,
   149, 153, 156, 163. *See also* Rose
   Cliff Hotel
Veblen, Thorstein, xiii, 35–36, 81, 85,
   136, 154. See also *Theory of the
   Leisure Class, The*
Vietnam, 159
Vigilantes, 21, 75–76

Wagner Act, 76
Wallace, Henry, 61–62, 74
Wallhausen, Art, 95, 121–22
Wallhausen, Millie, 103, 121, 123–24,
   139, 142–43
Warren, Julia Cooper, 139–42
Warrensburg, Mo., 26
Weapons of mass destruction, 132,
   137, 149
Whitfield, Barbara. *See* Fleming,
   Barbara Whitfield

Whitfield, Owen: and Snow, 5, 36,
   78–83, 89–90, 94, 100–101, 103–5,
   114, 116, 144, 155; and
   Sharecroppers' Roadside
   Demonstration, 6–7, 89–95,
   99–102, 135; as Baptist minister,
   36, 73, 114; and unionism, 73,
   79–81, 87, 95, 113–14; and threats
   against his family, 102, 105; work
   on behalf of sharecroppers,
   108–10; death of, 114
Whitfield, Shirley. *See* Farmer, Shirley
   Whitfield
Whitfield, Zella, 73, 87, 99, 102, 108,
   114
Williams, Mary Putnam, 14
Wilson, Woodrow, 129
Wolf Island, Mo., 12, 40, 49
Works Progress Administration
   (WPA), 50, 56–57, 95
World War I, 126, 128–29, 131–32
World War II, 110, 113, 125–27, 130,
   136, 148, 151, 159
Wright, Cleo, 123
Wyatt, Mo., ix, xiii, 98
Wyatt, William A., 42–43

Yarrow, Clarence H., 110

Zimmerman, Orville, 90–91